T0293636

Industrial Engineering Non-Traditional Applications in International Settings

Industrial Engineering: Management, Tools, and Applications

Industrial Engineering Non-Traditional Applications in International Settings

Industrial Engineering Applications in Emerging Countries

Global Logistics Management

Industrial Engineering Non-Traditional Applications in International Settings

Edited by

Bopaya Bidanda

İhsan Sabuncuoğlu

Bahar Y. Kara

CRC Press is an imprint of the
Taylor & Francis Group, an **informa** business

CRC Press
Taylor & Francis Group
6000 Broken Sound Parkway NW, Suite 300
Boca Raton, FL 33487-2742

CRC Press is an imprint of Taylor & Francis Group, an Informa business

Printed on acid-free paper
Version Date: 20140801

International Standard Book Number-13: 978-1-4822-2687-4 (Hardback)

Library of Congress Cataloging-in-Publication Data

Industrial engineering, non-traditional applications in international settings / editors, Bopaya Bidanda, Ihsan Sabuncuoglu, and Bahar Y. Kara.
pages cm
"A CRC title."
Includes bibliographical references and index.
ISBN 978-1-4822-2687-4 (alk. paper)
1. Industrial engineering. I. Bidanda, Bopaya. II. Sabuncuoglu Ihsan (Engineer) III. Kara, Bahar Y. (Bahar Yetis)

T56.I438 2014
670--dc23 2014021518

Visit the Taylor & Francis Web site at
http://www.taylorandfrancis.com

and the CRC Press Web site at
http://www.crcpress.com

This book is dedicated to my dear wife, Louella, and my children, Maya and Rahul, who have spent many evenings and weekends without a husband and a father, when I was in the pursuit of integrating the global industrial engineering profession.

Bopaya Bidanda

Contents

Preface

We are pleased to present to you this book that focuses on two key aspects of the industrial engineering (IE) discipline—*non-traditional settings* and *international environments*. IE was born and evolved in the United States during its few decades of existence. However, the last three decades have seen a dramatic shift in the application and evolution of this profession. The rapid growth we have witnessed in the IE profession and academic departments outside the United States has led to a significant body of IE knowledge being developed—this sometimes does not find appropriate archival outlets. We are pleased to provide such a forum. This book clearly illustrates how IE-based tools and techniques have been applied to a diversity of environments.

We therefore believe that this first-of-its-kind book can create an awareness of the breadth of the IE discipline both in terms of geography and scope.

Editors

Bopaya Bidanda is currently the Ernest E. Roth professor and chairman in the Department of Industrial Engineering at the University of Pittsburgh. His research focuses on manufacturing systems, reverse engineering, product development, and project management. He has published five books and more than a hundred papers in international journals and conference proceedings. His recent (edited) books include those published by Springer—*Virtual Prototyping & Bio-Manufacturing in Medical Applications* and *Bio-Materials and Prototyping Applications in Medicine*. He has also given invited and keynote speeches in Asia, South America, Africa, and Europe. He also helped initiate and institutionalize the engineering program on the Semester at Sea voyage in 2004.

He previously served as the president of the Council of Industrial Engineering Academic Department Heads (CIEADH) and also on the board of trustees of the Institute of Industrial Engineers. He also serves on the international advisory boards of universities in India and South America.

Dr. Bidanda is a fellow of the Institute of Industrial Engineers and is currently a commissioner with the Engineering Accreditation Commission of ABET. In 2004, he was appointed a Fulbright Senior Specialist by the J. William Fulbright Foreign Scholarship Board and the U.S. Department of State. He received the 2012 John Imhoff Award for Global Excellence in Industrial Engineering given by the American Society for Engineering Education. He also received the International Federation of Engineering Education Societies (IFEES) 2012 Award for Global Excellence in Engineering Education in Buenos Aires and also the 2013 Albert G. Holzman Distinguished Educator Award given by the Institute of Industrial Engineers. In recognition of his services to the engineering discipline, the medical community, and the University of Pittsburgh, he was honored with the 2014 Chancellors Distinguished Public Service Award.

İhsan Sabuncuoğlu is the founding rector of Abdullah Gul University. He earned his BS and MS in industrial engineering from the Middle East Technical University in 1982 and 1984, respectively. He earned his PhD in industrial engineering from Wichita State University in 1990.

Dr. Sabuncuoğlu worked for Boeing, Pizza Hut, and the National Institute of Heath in the United States during his PhD studies. He joined Bilkent University in 1990 and worked as a full-time faculty member until 2013. In the meantime, he held visiting positions at Carnegie Mellon University in the United States and at Institut Français de Mécanique Avancée (IFMA) in France. His research interests are in real-time scheduling, simulation optimization, and applications of quantitative methods to cancer-related health-care problems. His research has been funded by TUBITAK (The Scientific and Technological Research Council of Turkey) and EUREKA (a European-wide initiative to foster European competitiveness through cooperation among companies and research institutions in the field of advanced technologies).

Dr. Sabuncuoğlu also has significant industrial experience in aerospace, automotive, and military-based defense systems. His industrial projects are sponsored by a number of both national and international companies. He is currently the director of the Bilkent University Industry and the University Collaboration Center (USIM) and the chair of the Advanced Machinery and Manufacturing Group (MAKITEG) at TUBITAK.

In addition to publishing more than a hundred papers in international journals and conference proceedings, Dr. Sabuncuoğlu has edited two books. He is also on the editorial board of a number of scientific journals in the areas of industrial engineering and operations research. He is also a member of the Institute of Industrial Engineering, the Institute for Operations Research, the Management Sciences, and the Simulation Society. He is also a member of the Council of Industrial Engineering Academic Department Heads (CIEADH) and various other professional and social committees.

Bahar Y. Kara is an associate professor in the Department of Industrial Engineering at Bilkent University.

Dr. Kara earned an MS and a PhD from the Bilkent University Industrial Engineering Department, and she worked as a postdoctoral researcher at McGill University in Canada.

Dr. Kara was awarded Research Excellence in PhD Studies by INFORMS (Institute for Operations Research and Management Science) UPS-SOLA.

In 2008, Dr. Kara was awarded the TUBA-GEBIP (National Young Researchers Career Development Grant) Award. She attended the World Economic Forum in China in 2009. For her research and projects, the IAP (Inter Academy Panel) and the TWAS (The Academy of Science for the Developing World) Awarded her the IAPs Young Researchers Grant. Dr. Kara was elected as an associate member of Turkish Academy of Sciences in 2012. She has been acting as a reviewer for the top research journals within her field. Her current research interests include distribution logistics, humanitarian logistics, hub location and hub network design, and hazardous material logistics.

Contributors

Sercan Akkaş
Department of Industrial Engineering
İstanbul Technical University
İstanbul, Turkey

Mehmet A. Begen
Richard Ivey School of Business
University of Western Ontario
London, Ontario, Canada

Mehdi Bijari
Department of Industrial and System Engineering
Isfahan University of Technology
Ishahan, Iran

Mustafa Alp Ertem
Department of Industrial Engineering
Çankaya University
Ankara, Turkey

Emel Emur
Department of Industrial Engineering
Çankaya University
Ankara, Turkey

Ibrahim H. Garbie
Department of Mechanical and Industrial Engineering
College of Engineering
Sultan Qaboos University
Muscat, Oman

and

Department of Mechanical Engineering
College of Engineering at Helwan
Helwan University
Cairo, Egypt

Burcu Caglar Gencosman
Department of Industrial Engineering
Uludag University
Bursa, Turkey

Mahmut Ali Gökçe
Department of Industrial Engineering
Izmir University of Economics
Izmir, Turkey

Fazıl Gökgöz
Faculty of Political Sciences
Department of Management
Ankara University
Ankara, Turkey

Gül Işık
Department of Industrial Engineering
Izmir University of Economics
Izmir, Turkey

Mehdi Jafarian
Department of Industrial and System Engineering
Isfahan University of Technology
Ishahan, Iran

Seifedine Kadry
Department of Industrial Engineering
American University of the Middle East
Egaila, Kuwait

A. Argun Karacebey
Faculty of Economics and Business Administration
Department of Management
Okan University
İstanbul, Turkey

Erhan Kozan
School of Mathematical Sciences
Queensland University of Technology
Brisbane, Queensland, Australia

Shi Qiang Liu
School of Mathematical Sciences
Queensland University of Technology
Brisbane, Queensland, Australia

Amin Mousavi
School of Mathematical Sciences
Queensland University of Technology
Brisbane, Queensland, Australia

Kıvanç Onan
Department of Industrial Engineering
Doğuş University
İstanbul, Turkey

Erdinç Öner
Department of Industrial Engineering
Izmir University of Economics
Izmir, Turkey

Huseyin Ozkan
Beycelik Gestamp
Bursa, Turkey

H. Cenk Ozmutlu
Department of Industrial Engineering
Uludag University
Bursa, Turkey

Başar Öztayşi
Department of Industrial Engineering
İstanbul Technical University
İstanbul, Turkey

Ayşenur Sahin
Department of Industrial Engineering
Çankaya University
Ankara, Turkey

Ceren Salkın
Department of Industrial Engineering
İstanbul Technical University
İstanbul, Turkey

Bahar Sennaroğlu
Department of Industrial Engineering
Marmara University
İstanbul, Turkey

Füsun Ülengin
School of Management
Sabancı University
İstanbul, Turkey

Introduction

The discipline of industrial engineering has evolved for more than a hundred years. Over the last two decades, much of the growth of applications in industrial engineering has occurred outside the United States. This book focuses on non-traditional applications in international settings and will therefore detail some of the more exciting developments, applications, and implementations of industrial engineering and related tools. This book contains 10 chapters developed by authors and coauthors of at least six different countries. Though the chapters are arranged in no particular order, each represents a novel application of industrial tools and techniques.

Chapter 1 details an integrated model that can be applied to noncyclic maintenance planning, which can also be integrated with production planning. The authors develop a model that can be applied to smaller solutions, but since it is NP hard, they also call for more efficient heuristic models that can be developed and applied to large problems.

Chapter 2 presents a series of performance metrics for manufacturing (and possibly other types of) enterprises. These metrics are based on the degree or level of complexity, level of leanness, and level of agility. These metrics are then implemented and validated in an organization based in the Middle East.

Chapter 3 focuses on scheduling of automotive stamping operations, which are typically characterized by high set-up times and small unit run times. The authors present a solution utilizing mixed integer programming and apply this, in part, to the operations of a real live workplace. How this model can significantly increase production capacity by effective scheduling is also discussed.

Chapter 4 presents a completely new application in that Heider's balance theory is utilized to detect voting fraud in social media. This is especially important with the increasing utility of social media in our everyday lives.

Chapter 5 presents applications in the area of natural resource development, or more specifically open-pit mining. An optimization model is developed and applied (via case studies) to optimize the extraction sequence of blocks—an operation that can have a major impact on mining profitability.

Chapter 6 details where best to locate sites for disaster waste procession. Multiobjective optimization is used to identify site locations and provide solution guidance to this important, yet often unnoticed, problem that can occur at a moment's notice anywhere in our world.

Chapter 7 also studies disasters, but from a different perspective. It first details the shift in Turkey from crisis management to risk management and assigns disaster response facilities to near optimal locations. The work was driven by the risks posed by earthquake in Turkey and also in the locations where disaster response facilities currently exist.

Chapter 8 deals with a more pleasant topic—one that studies factors affecting buying patterns and behaviors at private shopping clubs. Turkey is taken as a benchmark and a technology acceptance model is used to study the buying behavior. Results from an already implemented questionnaire are also discussed.

Chapter 9 shifts gears to detail optimization methods that can be used to increase the effectiveness of the timing of traffic signals. With the rapid urbanization of emerging countries and related congestion in cities, this is a problem that will continue to receive much attention.

Chapter 10 discusses the Turkish banking sector and the measurement of efficiency of its banks, a topic that greatly impacts the emerging financial market. The authors apply quantitative models to study 29 commercial banks and 12 investment banks and show the relative efficiency (or lack thereof) of each type of bank.

As can be seen, there is a refreshing diversity of application environments and types of tools used.

1

Integrated Production Planning Model for Noncyclic Maintenance and Production Planning

MEHDI BIJARI AND MEHDI JAFARIAN

Contents

1.1 Introduction

Maintenance and production are closely related in different ways. This relationship makes production planning and maintenance planning the most important and demanding areas in the process of decision-making by industrial managers. Hence, they have been the focus of attention in the manufacturing industry while a lot of research has also been devoted to them in the area of operations research. Although the two activities are interdependent, they have most often been performed independently. Integration of production planning and maintenance planning into one single problem is a complex and challenging task since the resulting integrated planning problem leads to nonoptimal solutions.

Maintenance becomes necessary because of either a failure in production or the undesirably low quality of the items produced. However, the significance of maintenance planning can be more vividly realized when maximum plant availability and maximum mean time between equipment failures are sought at the lowest cost.

Maintenance activities may be classified into four types: corrective maintenance, predictive maintenance, repairs maintenance, and preventive maintenance (PM). Corrective maintenance can be defined as the maintenance that is required when an item has failed or worn out, to bring it back to working order. While predictive maintenance tends to include direct measurement of the item, repairs maintenance is simply doing maintenance work as need develops. This elementry approach has sometimes been replaced by periodic overhauls and other preventive maintenance activities. PM is performed periodically in order to reduce the incidence of equipment failure and the costs associated with it. These costs include disrupted production schedules, idled workers, loss of output, and damage to products or other equipment. PM, thus, improves production capacity, production quality, and overall efficiency of production plants. Moreover, it can be scheduled to avoid interference with production.

There are trade-offs between PM planning and production planning. PM activities take time that could otherwise be used for production, but delaying PM for production may increase the probability of machine failure. Whenever an unexpected machine failure occurs, the current production plan becomes inadequate and needs to be modified. Changes in production plan sometimes cause extra costs or significant changes in the service level and production line productivity.

Production planning mainly has two aspects: lot-sizing and scheduling. Lot-sizing concerns determining production quantity while scheduling concerns sequencing products on the production line. Decisions for these two problems are mostly made in a hierarchical manner. In other words, the lot-sizing problem is solved first and the output is used in the sequencing and scheduling problem. The problem is sometimes described as the general lot-sizing and scheduling problem; GLSP (Fleischmann and Meyr 1997) or capacitated lot-sizing problem with sequence-dependent setup times (CLSP-SD) that addresses the integrated lot-sizing and scheduling problems simultaneously due to their dependencies. In this chapter, a new integrated model is presented for the noncyclic maintenance and production planning problem. The Markov chain is used for producing the parameters required for processing the model of a single-stage multiparallel machine production system with the objective of maximizing profits with the assumption of demand flexibility. In this model, the value of maintenance has been taken into account. The product yield depends on equipment conditions, which deteriorate over time. The objective is to determine equipment maintenance schedule, demand quantity, and lot-sizes, and production schedules in such a way that the expected profit is maximized.

1.2 Literature Review

PM planning models are typically stochastic models accompanied by optimization techniques designed to maximize equipment availability or minimize equipment maintenance costs. There are mathematical or simulation models.

The literature abounds in papers on planning and optimizing maintenance activities. However, a few can be found dealing with models that combine PM planning and production planning.

The models reported in the literature include such decision variables as the number of maintenance activities, safety buffers, and inspection intervals. While the objective in most models is minimizing costs, some also consider system lifetime, which is generally assumed to have a Weibull distribution.

Models developed for integrating PM planning and production planning are Np-Hard, which can be optimally solved for small-size instances, but obtaining optimal solutions is impractical for large-size instances. This, therefore, warrants efficient solvers for large-size problems. Both heuristic and meta-heuristic methods including genetic algorithm, simulated annealing algorithm, and Lagrangian procedure, and expert systems have been used for the solution of these models.

Most papers in this area deal with production scheduling, and the models used for production scheduling and PM planning are designed with an implicit common goal of maximizing equipment productivity. Some studies extend the simple machine scheduling models by considering the maintenance decisions as given, or as constraints, rather than integrating them. The problem in these studies is modeled as a sequencing and scheduling problem with the machine availability constraint (Molaee et al. 2011).

Different methods have been used to develop models for the production planning and PM problems. Cassady and Kutanoglu (2005) have classified these methods into two broad approaches: reactive and robust. In the reactive approach, attempts are made to update production when a failure occurs. In robust planning, the plan is not sensitive to failure events. In another classification (Iravani and Duenyas 2005), the studies are classified into two groups. While the first focuses on the effect of failure on production schedule, the other group integrates production and maintenance planning into a single problem. Meller and Kim (1996) reviewed the literature and classified studies into two categories: one focusing on PM, and the other focusing on the statistical analysis of safety-stock-based failure neglecting PM.

Brandolese et al. (1996) considered a single-stage, multiproduct production environment with flexible parallel machines. They developed an expert system for the planning and management of a multiproduct and one-stage production system made up of flexible machines operating in parallel. The system schedules both production and maintenance at the same time.

Setup costs are sequence dependent. Sloan and Shanthikumar (2000) studied a multiproduct, single-machine problem in which the machine has states that change during the planning horizon such that the machine state affects the production rate of each product. They used the Markov chain and their objective function aimed to maximize profits. In each period, either a product is being produced or a maintenance activity is being performed. Their model determines the optimal policy of production and maintenance. The objective is achieving optimal maintenance policy in such a way that the sum of the discounted costs of maintenance, repairs, production, backorders, and inventory is minimized.

Aghezzaf and Najid (2008) presented a production plan and a maintenance plan in a multiproduct, parallel machine system with corrective maintenance and PM. It is assumed that when a production line fails, a minimal repair is carried out to restore it to an *as-bad-as-old* status. PM is also carried out periodically at the discretion of the decision maker to restore the production line to an *as-good-as-new* status. The resulting integrated production and maintenance planning problem is modeled as a nonlinear mixed-integer program when each production line implements a cyclic PM policy. When noncyclic PM policies are allowed, the problem is modeled as a linear mixed-integer program. In this situation, maintenance activities decrease production capacity. Sitompul and Aghezzaf (2011) proposed an integrated production and maintenance hierarchical plan. Noncyclic maintenance in the single machine problem has been considered in another study (Nourelfath et al. 2010), in which it is assumed that while production capacity is constant, a decision must be made in each period about implementing the PM.

Fitouhi and Nourelfath (2012) extended upon previous studies (Nourelfath et al. 2010). They proposed a model in which the noncyclic maintenance assumption was abandoned and the assumption that the machine has several states due to its components was adopted instead. The proposed model coordinates the production with the maintenance decisions so that the total expected cost is minimized. We are given a set of products that must be produced in lots on a multistate production system during a specified finite planning horizon. Planned PM and unplanned corrective maintenance can be performed on each component of the multistate system. The maintenance policy suggests cyclic preventive replacements of components and a minimal repair on failing components. The objective is to determine an integrated lot-sizing and PM strategy of the system that will minimize the sum of preventive and corrective maintenance costs, setup costs, holding costs, backorder costs, and production costs, while satisfying the demand for all products over the entire horizon. Production yield is influenced by the machine state.

Yao et al. (2005) studied the joint PM and production policies for an unreliable production-inventory system in which maintenance/repair times are nonnegligible and stochastic. A joint policy decides (1) whether or not to perform PM and (2) if PM is not performed, then how much to produce. A discrete-time system is considered, and the problem is formulated as a Markov decision process model. Although their analysis indicates that the structure of the optimal joint policies is generally very complex, they were able to characterize several properties regarding PM and production including optimal production/maintenance actions under backlogging and high inventory levels. Wee and Widyadana (2011) studied the economic production quantity models for deteriorating items with rework and stochastic PM time.

Lu et al. (2013) studied system reliability. According to them, a system reliability lower bound is determined that is smaller than the system reliability. Marais and Saleh (2009) investigated the maintenance value and proposed that maintenance has an intrinsic value. They argue that the existing cost-oriented models ignore an important dimension of maintenance activities that involves quantifying their value. They consider systems that deteriorate stochastically and exhibit multistate failures. The state evolution is modeled in their study using the Markov chain and directed graphs. To account for maintenance value, they calculate the net present value of maintenance in their model.

Njike et al. (2011) used the value-optimized concept in their research. They sought to develop an optimal stochastic control model in which interactive feedback consisted of the quantity of flawless and defective products. The main objective was to minimize the expected discounted overall cost due to maintenance activities, inventory holding, and backlogs. A near-optimal control policy of the system was then obtained through numerical techniques. The originality of their research lies in the fact that all operational failures have been taken into account in the same optimization model. This brings a value added to the high level of maintenance and for operation managers who need to consider all failure parameters before taking cost-related decisions.

1.3 Integrated Model for Noncyclic Maintenance Planning and Production Planning

1.3.1 Problem Statement

In this section, an integrated model is presented for the noncyclic maintenance planning and production planning problem. The objective is to maximize profits. The model considers simultaneous lot-sizing and scheduling. The challenge commonly faced within production planning is the coordination of demand and production capacity.

Demand flexibility assumption is also introduced into the model. In many industries, product yield is heavily influenced by equipment conditions. Previous studies have focused either on maintenance at the expense of the effect of equipment conditions on yield or on production at the expense of the possibility for actively changing machine state.

1.3.2 *Assumptions*

The assumptions made here are classified into those related to production planning and those concerning PM planning.

Assumptions of production planning:

- The model is a multiproduct one.
- The model is in a multiparallel machine environment.
- The available capacity is finite. Considering PM planning in each period, the capacity may take different values as computed based on mathematical expectation.
- The planning horizon is finite and consists of T periods.
- The demand for a product is not known before each period and is determined by the model. For each product, this value ranges between a lower and an upper bound for the demand.
- Shortage is allowed in periods. However, the total demand should be met at the end of the planning horizon.
- Setup times and setup costs are sequence dependent.
- Holding costs, setup costs, and production costs are time independent.
- The model has the characteristic of setup preservation. It means that if we have an idle time, the setup state would not change after it.
- Lots are continuous. This means that production can continue for the next period with no break and with no setup.
- The setup state is specific at the beginning of the planning horizon.
- It is possible to produce some types of products in each period. In other words, the model is a big-time bucket one.
- The objective function is maximizing the sales revenue minus production, holding, and setup costs.
- The breakdown of setup time between two periods is not allowed, and the setup is finished in the same period in which it begins.

Assumptions of PM planning:

- The machine has the following three states:
  1. Working at good efficiency (state 1).
  2. Working at low efficiency (state 2).
  3. Where the machine breaks down, a non-PM repair state starts after a sudden breakdown (state r).
- Product quality is not influenced by the machine state.
- Maintenance operation does not create a disturbance or a change in the setup state.
- The PM operation is an activity with a positive effect; it increases system efficiency. There is at least one state in which the PM improves system efficiency.
- For PM planning, microperiods are considered to be separate from microperiods in production planning. Therefore, the PM schedule is discrete.
- In each microperiod, it is decided whether one and only one PM is to be performed or not.
- The efficiency or the capacity of a machine is reduced by production or as a result of exploiting this capacity. Therefore, the state of the machine goes toward state 2 or state r.
- The state of the machine turns to state 1 after a PM operation.
- Both the PM operation and the emergency maintenance operation are costly. In addition, they reduce the capacity of the period as they use this capacity.
- Only one PM operation is possible in each maintenance microperiod. On the other hand, it is assumed that only one sudden breakdown may happen in each microperiod.
- Transition of machine state is memory less from one period to the next.

In addition to these assumptions, the proposed model considers the existence of triangular inequality conditions or their paucity. In most industries, setup times conform to the triangular inequality. This assumption, which can also be applied to costs, is stated as follows:

$$Sc_{ik} + Sc_{kj} > Sc_{ij} \tag{1.1}$$

where Sc_{ij} represents the setup time from product i to product j.

In simple words, the triangular inequality states that the setup time or the setup cost for moving directly from product *i* to product *j* is less than the time or cost when there is a mediator.

In some industries, it is plausible that setup times do not conform to the triangular inequality. Changing color in some industries can be mentioned as an example; changing color from black to white needs more setup time than changing color from black to blue then from blue to white.

1.3.3 Profit Maximization of General Lot-Sizing and Scheduling Problem

The proposed integrated model for maintenance planning and production planning is based on the profit maximization general lot-sizing and scheduling problem (PGLSP) model (Sereshti 2010). Recently, attention has been directed toward the simultaneous lot-sizing and scheduling problem that has come to be called the general lot-sizing and scheduling problem or GLSP. For modeling this problem, two distinctive approaches may be employed. In the first approach, there are two kinds of time buckets, small buckets and large ones. Small buckets or positions are within the large buckets or macroperiods. The positions, or microperiods, are used for sequencing. This approach was first presented by Fleischmann and Meyr (1997). Meyr (2000) extended GLSP to deal with sequence-dependent setup times. The second approach is based on the CLSP-SD, which is related to the traveling salesman problem (Almada-Lobo et al. 2008).

The profit maximization of GLSP with demand choice flexibility is an extension of the GLSP in which the assumption of flexibility in choosing demands is also included. The accepted demand in each period can vary between its upper and lower bounds. The upper bound could be the forecasted demand, and the lower bound is the organization commitments toward customers or minimum production level according to production policy. PGLSP can be described as follows.

Having P products and T planning periods, the decision maker seeks to determine (1) the accepted demand of each product in each period, which is between an upper and a lower bound, (2) the quantity of lots for each product, and (3) the sequence of lots. The objective

Table 1.1 Parameters

T	Number of planning periods	
P	Number of products	
N	Number of positions in planning horizon	
π_n	The period in which position n is located	$n = 1,\ldots,N$
C_t	Available capacity in each period	$t = 1,\ldots,T$
Ld_{jt}	Demand lower bound for product j in period t	$j = 1,\ldots,P, t = 1,\ldots,T$
Ud_{jt}	Demand upper bound for product j in period t	$j = 1,\ldots,P, t = 1,\ldots,T$
n_t	Number of positions in period t	$t = 1,\ldots,T$
F_t	First position in period t	$t = 1,\ldots,T$
L_t	Last position in period t	$t = 1,\ldots,T$
h_j	Holding cost for one unit of product j	$j = 1,\ldots,P$
r_{jt}	Sales revenue for one unit of product j in period t	$j = 1,\ldots,P, t = 1,\ldots,T$
Cp_j	Production cost for one unit of product j	$j = 1,\ldots,P$
p_j	Processing time for one unit of product j	$j = 1,\ldots,P$
S_{ij}	Setup cost for transition from product i to product j	$i = j = 1,\ldots,P$
St_{ij}	Setup time for transition from product i to product j	$i = j = 1,\ldots,P$
I_{j0}	Initial inventory level for product j	$j = 1,\ldots,P$

function is maximizing the sales revenues minus production, holding, and setup costs. Backlog is not allowed. Setup times and costs are sequence dependent. The triangular inequality lies between setup times. Back order is not allowed.

The parameters for this model are presented in Table 1.1.

This model is an extension of the model proposed by Meyr (2000). PGLSP has also been modeled through the traveling salesman problem approach (Sereshti and Bijari 2013). In Meyr's model, the microperiods or positions within the planning periods are used as a modeling consideration to define the sequence of products. The number of these microperiods in each macroperiod forms a parameter of the model, and they are used to define the first and last positions in each period. The decision variables for this model are as follows:

I_{ij} = Inventory level of product j at the end of period t.

D_{jt} = Accepted demand of product j in period t.

Q_{jt} = Quantity of product j produced in position n.

Y_{jt} = A binary variable that is 1 when the setup state in position n is for product j.

X_{ijn} = A positive variable whose amount is always 0 or 1. This variable is 1 when the setup state changes from product i to product j in position n.

The mathematical model is presented as follows:

$$\text{Max} \sum_{j=1}^{P}\sum_{t=1}^{T} r_{jt} D_{jt} - \sum_{j=1}^{P}\sum_{n=1}^{N} Cp_j Q_{jn} - \sum_{j=1}^{P}\sum_{i=1}^{P}\sum_{n=1}^{N} S_{ij} X_{ijn} - \sum_{j=1}^{P}\sum_{t=1}^{T} h_j I_{jt}$$

subject to

$$I_{jt} = I_{j(t-1)} + \sum_{n=F_t}^{L_t} Q_{jn} - D_{jt} \quad j = 1,\ldots,P, \quad t = 1,\ldots,T \tag{1.2}$$

$$Ld_{jt} \le D_{jt} \le Ud_{jt} \quad j = 1,\ldots,P, \quad t = 1,\ldots,T \tag{1.3}$$

$$Q_{jn} \le M_{j\pi_n} Y_{jn} \quad j = 1,\ldots,P, \quad n = 1,\ldots,N \tag{1.4}$$

Other constraints of the model are the same as the Meyr's model.

The objective function of model is to maximize sales revenues minus production, setup, and holding costs. Constraint (1.2) shows the balance among demand, production, and inventory. Constraint (1.3) guarantees that the accepted demand for each product in each period is between its upper and lower bounds. Constraint (1.4) ensures that a product can be produced when its setup is complete. The upper bound of production in this constraint can be seen in statement (1.5). If we just use c_t/p_j as the upper bound, the constraint will be true; using the maximum value of the remaining demand in the following period may result in a tighter constraint, which occurs when the remaining demand is less than the production capacity.

$$M_{jt} = \min\left\{\frac{C_t}{p_j}, \sum_{k=t}^{T} Ud_{jk}\right\} \quad j = 1,\ldots,P, \quad n = 1,\ldots,N \tag{1.5}$$

1.3.4 Integrated Model

We have used PGLSP for modeling our problem (Bijari and Jafarian 2013). In the model, both macroperiods and microperiods are considered as in the basic GLSP. The decision maker wants to define the mentioned decisions in the previous section and also the period in which PM will be executed. The objective function is maximizing sales revenues minus production, holding, shortage costs, setup, PM, and non-PM costs.

The model is stochastic because the machine state is stochastic, too. The objective is maximizing the expected value of profits. Machine has three states. We use production microperiods for producing products. Maintenance microperiod is used for performing maintenance activities. In each period, the probability of machine state after the last PM can be determined. The parameters of the production microperiods are as follows:

R: Number of products
π_n: Number of microperiods containing position n
n_t: Number of positions available in period t
nr_t: Number of maintenance microperiods in period t
β: Coefficient of efficiency when the machine is in state 2
Cp_j: Production cost of j
ρ_{jk}: Usage rate of machine k for producing item j
S_{ijk}: Setup cost of machine k for producing item j after item i
St_{ijk}: Setup time of machine k for producing item j after item i
e: Discount rate in each period
b_j: Shortage cost in each period
cr: Non-PM cost
m_j: Minimum production lot-size of j
$P_i^{\tau r}$: Probability of state 1 after $(\tau r - 1)$ microperiods from last PM
π_{nr}: Number of macroperiods that include maintenance microperiod nr
Crp: PM cost
U_k: Maintenance microperiod capacity for machine k
δ_{ij}: Probability of transition from state i to j

The model and its decision variables are proposed as follows:

I_{jt}^+: Inventory of product j at the end of period t
I_{jt}^-: Shortage of product j at the end of period t
Q_{jnk}: Production quantity of product j in microperiod n for machine k

tr_{nrk}: The number of microperiods (plus 1) between the last PM and the maintenance period nr for machine k

Y_{jnk}: Binary variable; it is 1 if product j is produced in position n on machine k

X_{ijnk}: If machine K setup is accomplished for producing product j in position n; when product i is produced in this position, it is 1; otherwise, zero $j, i = 1, \ldots, R$, $k = 1, \ldots, K$, $n = 1, \ldots, N$

q_k^{nr}: Binary variable; PM was done (1) or was not done in the maintenance microperiod nr on machine k

Other parameters and decision variables are the same as PGLSP. Production microperiods parameters are expressed as follows:

F_{tk}: The first position in period t for machine k, $k = 1, \ldots, K$, $t = 1, \ldots, T$

L_{tk}: The last position in period t for machine k

N: Total number of positions at planning horizon

$$F_t = \sum_{k=1}^{t-1} n_k + 1 \tag{1.6}$$

$$L_t = F_t + n_t - 1 \tag{1.7}$$

$$N = \sum_{t=1}^{T} n_t \tag{1.8}$$

Maintenance (M) microperiods parameters are as follows:

F_{rtk}: The first PM position in period t for machine k

L_{rtk}: The first PM position in period t for machine k

Nr: Total number of PM positions

$$F_{rt} = \sum_{k=1}^{t-1} nr_k + 1 \tag{1.9}$$

$$L_{rt} = F_{rt} + nr_t - 1 \tag{1.10}$$

$$Nr = \sum_{t=1}^{T} nr_t \tag{1.11}$$

The model is shown as follows:

$$\max E(Z) = \sum_{j=1}^{R}\sum_{t=1}^{T}(1-e)^{t-1} r_{jt} D_{jt} - \sum_{j=1}^{R}\sum_{n=1}^{N}\sum_{k=1}^{K}(1-e)^{t-1} cp_j Q_{jnk}$$

$$-\sum_{j=1}^{R}\sum_{i=1}^{R}\sum_{n=1}^{N}\sum_{k=1}^{K}(1-e)^{\pi_n -1} S_{ij} X_{jink} - \sum_{j=1}^{R}\sum_{t=1}^{T}(1-e)^{t-1} h_j I_{jt}^{+}$$

$$-\sum_{j=1}^{R}\sum_{t=1}^{T}(1-e)^{t-1} b_j I_{jt}^{-} - cr\sum_{k=1}^{K}\sum_{nr=1}^{Nr}\sum_{\tau r}^{Nr}\frac{(1-e)^{\pi_{nr}-1} Pr^{\tau r}}{\left(tr_{nrk} - tr\right)M + 1}$$

$$-crp\sum_{k=1}^{K}\sum_{nr=1}^{Nr}(1-e)^{\pi_{nr}-1} q_k^{nr} \tag{1.12}$$

subject to

$$I_{jt}^{+} = I_{j(t-1)}^{+} - I_{j(t-1)}^{-} + \sum_{k=1}^{K}\sum_{n=F_{tk}}^{L_{tk}} Q_{jnk} - D_{jt} + I_{jt}^{-} \quad \forall t, j \tag{1.13}$$

$$\sum_{k=1}^{K}\sum_{n=1}^{N} Q_{jnk} = \sum_{t=1}^{T} D_{jt} \quad \forall j \tag{1.14}$$

$$Ld_{jt} \le D_{jt} \le Ud_{jt} \quad \forall t, j \tag{1.15}$$

$$Q_{jnk} \le MY_{jnk} \quad \forall j, n, k \tag{1.16}$$

$$\sum_{j=1}^{R}\sum_{n=F_{tk}}^{L_{tk}} \rho_{jk} Q_{jnk} + \sum_{i=1}^{R}\sum_{j=1}^{R}\sum_{n=F_{tk}}^{L_{tk}} St_{ijk} X_{ijnk}$$

$$\le \sum_{nr=F_{rtk}}^{L_{rtk}}\sum_{\tau r=1}^{Nr}\frac{\left[P_1^{\tau r} U_k + P_2^{\tau r} U_k^{\beta} \dagger\right]}{\left(tr_{nrk} - \tau r\right)M + 1} \quad \forall k, t \tag{1.17}$$

$$\sum_{j=1}^{R} Y_{jnk} = 1 \quad \forall n, k \tag{1.18}$$

$$X_{ijnk} \ge Y_{i(n-1)k} + Y_{jnk} - 1 \quad \forall j, i, n, k \tag{1.19}$$

Table 1.2 Transition Matrix

	1	2	R
1	δ_{11}	δ_{12}	δ_{1r}
2	0	δ_{22}	δ_{2r}
3	1	0	0

$$tr_{nrk} = tr_{(nr-1)k}\left(1-q_k^{nr}\right)+1 \quad \forall n,k \tag{1.20}$$

$$Y_{jnk},\quad q_k^{nr} \in \{0,1\} \quad \forall j,n,k,nr \tag{1.21}$$

$$X_{ijnk}, Q_{jnk}, I_{jt}^{+}, I_{jt}^{-}, tr_{nrk}, D_{jt} \geq 0 \quad \forall j,n,k,i \tag{1.22}$$

The transition matrix is shown in Table 1.2.

The probability of state i after $(\tau r - 1)$ microperiods from last PM can be obtained by the following equations:

$$P_1^{tr} = P_1^{tr-1}\delta_{11} + P_r^{(tr-1)}$$

$$P_2^{tr} = P_2^{tr-1}\delta_{22} + P_1^{tr-1}\delta_{12}$$

$$P_r^{tr} = P_2^{tr-1}\delta_{2r} + P_1^{tr-1}\delta_{1r}$$

$$P_1^1 = P_2^1 = P_r^1 = 0$$

$$P_1^2 = 1,\quad P_2^2 = P_r^2 = 0$$

All P_i^1 $(tr = 1)$ are equal to zero when PM is performed on the machine.

A discount rate is used in the objective function. It designates the value of maintenance. The first term in the objective function is related to sales revenue. Other terms designate production cost, setup cost, holding cost, shortage cost, expected value of non-PM cost, and PM cost. Non-PM cost is determined by multiplying the emergency non-PM cost by the probability of this state ($P_r^{\tau r}$) after $\tau r - 1$ from the last PM period. The denominator ensures that the value of τr is properly chosen. Only when tr equals tr_{nrk}, the denominator equals 1; otherwise, it has a big value because M is a big value. Thus, fractions become zero.

Constraint (1.13) shows production, demand, inventory, and shortage balance. Constraint (1.14) ensures that production quantities are equal to the satisfied demand. Constraint (1.15) shows the demand range. The next constraint shows the relation between setup and production feasibility. Constraint (1.17) ensures that the machine usage for production and setup has not exceeded the available capacity. The right side of this constraint estimates the available capacity. The numerator calculates the expected value of capacity. $P_r^{\tau r}\mu$ modifies the error of unequal. The denominator ensures that the value of τr is properly chosen. Constraint (1.18) shows that only one product can be produced in each microperiod. Constraint (1.19) ensures that if two different products are manufactured in two consecutive microperiods, then setup will be necessary. Constraint (1.20) counts the number of periods since the last PM. As long as PM is not performed, that is, $q_k^{nr} = 1$, the value of tr_{nrk} per each maintenance microperiod is one unit greater than that of the previous maintenance microperiod; otherwise, its value is only 1, which means that PM occurred in the maintenance microperiod nr. Constraint (1.23) also limits the minimum batch size production. The constraint is added for considering the minimum batch size (m_j), in each machine setup. It can be written as follows:

$$Q_{jnk} \geq m_j \times (Y_{jnk} - Y_{j(n-1)k}) \quad \forall j, n, k \tag{1.23}$$

In some industries, if setup occurs, the production batch size must then be greater than a minimum level due to technological or economic factors. This constraint ensures that the minimum batch size is produced after each setup. It ensures that if the setup for a product was not carried out in microperiod (position) $n - 1$ but that it was in position n, then the product batch size in period t must be at least equal to the minimum batch size of the product.

1.3.5 Model Output Representation

The output of model should contain the production schedule and the PM schedule. The maximum number of lots equals the number of positions in the model. Therefore, the number of lots may be less than the number of positions. In this state, setup carryover is applied to the remaining positions at the end of the period. In other words, one setup state in each position is determined in the yield solution while production may plausibly not occur in some positions at the end of the period. The following conditions may be regarded as an example.

There are three types of products and five microperiods in each period. This means five lots can be produced in each period. If it is assumed that the triangular inequality conditions do not hold between setup times, there might be two or more product lots in one period. However, assuming that triangular conditions hold between setup times, production of a product occurs only in one lot in each period. Therefore, it will not be necessary for the number of microperiods in each period to exceed the number of product types. In addition, it is worth mentioning that if there are three types of products, there will be no need for the number of microperiods to be greater than the number of products under any assumption. For this problem size, there is no need for creating a complex state with more microperiods. However, as the number of products increases, problem complexity may also increase to the extent that prediction of the number of microperiods becomes impossible, especially when we are simultaneously faced with setup time and setup cost.

1.3.6 Hybrid Solution Algorithm

Given the fact that the PGLSP is NP-hard, the model presented in this chapter is NP-hard, too. Hence, efficient methods need to be developed that can obtain near-optimal solutions in a reasonable time for large-size instances. A simulating annealing (SA) algorithm and a hybrid algorithm have been developed for this purpose. The hybrid algorithm combines a heuristic algorithm and SA. The heuristic algorithm has two parts. Part 1 satisfies the minimum product demand. Part 2 assigns available capacities to products with higher profits. SA determines the product sequence and the PM schedule. Lot-sizing and demand quantity are obtained by the heuristic algorithm. The solutions obtained from solving the mathematical models have been used to assess the quality of the algorithms. Numerical results show the efficiency of the developed hybrid algorithm.

Acknowledgment

The authors thank Dr. Ezzatollah Roustazadeh from Isfahan University of Technology for editing the final English manuscript of this chapter.

References

Aghezzaf, E.H. and N.M. Najid. 2008. Integrated production planning and preventive maintenance in deteriorating production systems. *Information Sciences* 178: 3382–3392.

Almada-Lobo, B., D. Klabjan, M.A. Carravilla, and J.F. Oliveira. 2007. Single machine multi-product capacitated lot sizing with sequence-dependent setups. *International Journal of Production Research* 45: 4873–4894.

Bijari, M. and M. Jafarian. 2013. An integrated model for non cyclical maintenance planning and production planning. *Proceedings of International IIE Conference*, Istanbul, Turkey.

Brandolese, M., M. Franci, and A. Pozzetti. 1996. Production and maintenance integrated planning. *International Journal of Production Research* 34: 2059–2075.

Cassady, C.R. and E. Kutanoglu. 2005. Integrating preventive maintenance planning and production scheduling for a single machine. *IEEE Transactions on Reliability* 54: 304–309.

Fitouhi, M.C. and M. Nourelfath. 2012. Integrating noncyclical preventive maintenance scheduling and production planning for a single machine. *International Journal of Production Economics* 136: 344–351.

Fleischmann, B. and H. Meyr. 1997. The general lot-sizing and scheduling problem. *OR Spectrum* 19: 11–21.

Iravani, S.M.R. and I. Duenyas. 2002. Integrated maintenance and production control of a deteriorating production system. *IIE Transactions* 34: 423–435.

Lu, Z., Y. Zhang, and X. Han. 2013. Integrating run-based preventive maintenance into the capacitated lot sizing problem with reliability constraint. *International Journal of Production Research* 51: 1379–1391.

Marais, K.B. and J.H. Saleh. 2009. Beyond its cost, the value of maintenance: An analytical framework for capturing its net present value. *Reliability Engineering System Safety* 94: 644–657.

Meller, R.D. and D.S. Kim. 1996. The impact of preventive maintenance on system cost and buffer size. *European Journal of Operational Research* 95: 577–591.

Meyr, H. 2000. Simultaneous lot-sizing and scheduling by combining local search with dual reoptimization. *European Journal of Operational Research* 120: 311–326.

Molaee, E., G. Moslehi, and M. Reisi. 2011. Minimizing maximum earliness and number of tardy jobs in the single machine scheduling problem with availability constraint. *Computers & Mathematics with Applications* 62: 3622–3641.

Njike, A., R. Pellerin, and J.P. Kenne. 2011. Maintenance/production planning with interactive feedback of product quality. *Journal of Quality in Maintenance Engineering* 17: 281–298.

Nourelfath, M., M. Fitouhi, and M. Machani. 2010. An integrated model for production and preventive maintenance planning in multi-state systems. *IEEE Transactions on Reliability* 59: 496–506.

Sereshti, N. 2010. Profit maximization in simultaneous lot-sizing and scheduling problem. MSc dissertation, Isfahan University of Technology, Isfahan, Iran.

Sereshti, N. and M. Bijari. 2013. Maximization in simultaneous lot-sizing and scheduling problem. *Applied Mathematical Modelling* 37: 9516–9523.

Sitompul, C. and E.H. Aghezzaf. 2011. An integrated hierarchical production and maintenance-planning model. *Journal of Quality in Maintenance Engineering* 17: 299–314.

Sloan, T.W. and J.G. Shanthikumar. 2000. Combined production and maintenance scheduling for a multiple-product, single-machine production system. *Production and Operations Management* 9: 379–399.

Wee, H.M. and G.A. Widyadana. 2011. Economic production quantity models for deteriorating items with rework and stochastic preventive maintenance time. *International Journal of Production Research* 35: 1–13.

Yao, X., X. Xie, M.C. Fu, and S.I. Marcus. 2005. Optimal joint preventive maintenance and production policies. *Naval Research Logistics* 52: 668–681.

2

Non-Traditional Performance Evaluation of Manufacturing Enterprises

IBRAHIM H. GARBIE

Contents

2.1 Introduction and Motivation

Non-traditional evaluation of manufacturing enterprises regarding the existing status will be suggested and discussed in this chapter. There are many nonconventional aspects for measuring performance measurements in the manufacturing organizations/firms. These aspects are represented into level of complexity, level of leanness, and level of agility. These aspects are also considered as performance measurements. In this chapter, these performance measurements are used as a new evaluation of the existing status of manufacturing enterprise or firms. With respect to complexity, manufacturing firms require reduction in their complexity. Complexity in manufacturing firms presents a new challenge, especially, during the existing global recession. Estimating the level of complexity in manufacturing firms is still unclear due to difficulty of analysis. Lean thinking and/or lean manufacturing, which is mainly focusing on minimizing the wastes in production processes, is considered the second aspect of non-traditional evaluation. Measuring the level of manufacturing leanness is most important especially when it is considered as one of the most important strategies in manufacturing firms to increase their utilization of resources, processes, and materials. The last performance measure concerns agility, where manufacturing firms have great interest in developing their manufacturing systems to be more competitive in terms of flexibility and capability. Agile philosophy will be considered as one important issue of the next industrial revolution as a core prerequest of sustainable manufacturing enterprises. It is a manufacturing and/or management strategy that integrates technology, people, production strategies, and industrial organization management systems. In this chapter, three proposals for estimating complexity, leanness, and agility were suggested and discussed. A fuzzy logic approach was proposed to estimate these levels of the manufacturing firms. An illustrative example is used to obtain a very clear understanding of complexity, leanness, and agility.

2.2 Importance and Background

The complexity in any organization has a direct impact on its performance. Reducing the complexity in industrial/service organizations reduces their costs and also increases their revenue and enhances their competence in local and international markets. Complexity has a direct relationship between inputs and outputs of the organization. As organizations grow bigger and expand to satisfy their demand, they tend to have more complex supply management and manufacturing operations than simple ones. Today, several definitions of complexity exist as immense international interest and knowledge for the scientific basis. First, industrial and/or manufacturing complexity was defined as systemic characteristics that integrate several key dimensions of the manufacturing environment including size, variety, information, uncertainty, control, cost, and value (Garbie and Shikdar, 2011a,b). Also, flexibility and agility are considered as the most desirable of certain system properties for the manufacturing enterprises with respect to structural and operational complexity measures. These properties will give industrial organizations more ability to cope with increased environmental uncertainty and adapting to the faster pace of change of today's markets (Giachetti et al., 2003).

There are two different forms of complexity: (1) static or structural complexity, which is designed into the system architecture, (2) operational or dynamical complexity, which also can change dramatically in short periods of time according to its environment. Although most measurements were concentrated on operational measures, both structural and operational characteristics are important for the performance of the system as a whole. Determining the industrial system complexity still has different concepts and views. Also complexity in manufacturing systems was divided into two different categories: time-independent complexity and time-dependent complexity (Kuzgunkaya and ElMaraghy, 2006). Time-independent complexity is used to add the complexities arising from the designer's perception while time-dependent complexity is either combinational or periodic. The structural complexity measure is very close to time-independent complexity and provides a good description of the inherent complexity of its components, the relationship among them, and their influence (Kuzgunkaya and ElMaraghy, 2006). But dynamic complexity is more applicable to the system time-dependent behavior and requires data normally obtained during actual operations or simulation of the shop floor (Garbie, 2012a,b).

Reducing complexity level is a key factor for reducing costs and enhancing operating performance in many organizations. The more reduction in complexity in the organization, the greater is the customer expectations. This will lead to improve system's reliability, find out the particular parts of complexity of an organization, and measure overall performance (Garbie and Shikdar, 2011a). Optimizing the complexity in industrial organizations was recommended to be one of several solutions for the recovery of the existing financial recession (Garbie, 2009, 2010; Garbie and Shikdar, 2011b). Structural complexity on job shop manufacturing system was investigated considering processing time and scheduling rules (Jenab and Liu, 2010). There are also many types of complexities mentioned by several academicians such as process complexity and operational complexity. While the process complexity analysis focuses on the tools, equipment, and operations used to manufacture it (Hu et al., 2008b), operational complexity was considered as the cognitive and physical effort associated with the tasks related to a product/process combination. Complexities in supply chain management are considered as complexity issues regarding manufacturing enterprises such as upstream complexity, internal manufacturing complexity, and downstream complexity (Bozarth et al., 2009). Measuring the manufacturing complexity in assembly lines based on assembly activities is presented with different configurations and manufacturing strategies (Wu et al., 2007). Also, the effect of scheduling rules with processing times on hybrid flow a system is investigated (Yang, 2010). The complexity levels in industrial firms are estimated through several case studies based on a general framework that includes a questionnaire focusing on each issue in a firm (Garbie and Shikdar, 2011a).

They concluded that complexity arises from not only the size of the system but also the interrelationships of the system components and the emergent behavior that could not be predicted from the individual system components (Cho et al., 2009). Also, complexity can be classified into four different types: time-independent real complexity, time-independent imaginary complexity, time-dependent combinatorial complexity, and time-dependent periodic complexity. Also, technological complexity can be considered as another type of manufacturing complexity analysis (Tani and Cimatti, 2008) especially when applied to engineering and industrial manufacturing. Analysis of complexity is widely used to analyze the industrial enterprises or firms (Garbie and Shikdar, 2011a) as it is considered as one of important issues to reconfiguring manufacturing enterprises (Garbie, 2013a) and for sustainability (Garbie, 2013b).

Since 1980s until 1990s, manufacturing analysts have used the terms *lean production/manufacturing* for achieving greater flexibility, optimizing inventory, minimizing manufacturing lead-times, and increasing the level of quality in both products and customer service. The lean manufacturing is defined as a systematic approach to identifying and eliminating wastes or non-value-added activities through continuous improvement by flow of the product(s) at the pull of the customer in purist of perfection (Thomas et al., 2012). It can be expressed in industrial/manufacturing firms as the performance-based process to increase competitive advantage. The basics of lean manufacturing employ continuous improvement processes in order to focus on the elimination of wastes or non-value-added activities within an organization. The challenge to organizations utilizing lean manufacturing is to create a culture that will create and sustain long-term commitment. Toyota production system is considered the leading lean exemplar in the world. It became the largest car maker in the world in terms of overall sales due to adopting lean thinking.

Lean thinking brings growth to every manufacturing company in the world year after year as a new manufacturing and/or management philosophy to maximize productivity and quality, and minimizing costs. The managers are also adapting to the tools and principles beyond manufacturing in different areas such as logistics and distribution, services, retail, healthcare, construction, maintenance, and even government. Therefore, lean thinking is beginning to implement its tools and techniques in all sectors today in general and in manufacturing sector in specific. Lean and six sigma are used as subgoals to measure the performance measurement in manufacturing companies (Hu et al. 2008a).

In 1991, about two decades ago, when the industry leaders were trying to formulate a new paradigm for successful manufacturing enterprises in the twenty-first century, even though many manufacturing firms were still struggling to implement lean thinking and concepts, the agile manufacturing paradigm was formulated in response to the constantly changing *new economy* as a basis for returning to global competitiveness based on practical study under the auspices of the Iacocca Institute at Lehigh University. This study was sponsored by the US Navy Mantech program and involved 13 US companies. The objective of the study was to consider what the characteristics would be that successful manufacturing companies will possess (Groover, 2001). By the time the study was completed, more than 100 companies had participated in addition to the original 13 companies. The report of the study was entitled "21st Century Manufacturing Enterprise Study." The term *agile manufacturing* was coined to describe a new manufacturing paradigm that was recognized as emerging to replace mass production. Agility means different things to different enterprises under different contexts. Agility is characterized by cooperativeness and synergism, a strategic vision, responsive creation and customer-valued delivery, nimble organization structures, and an information infrastructure

(Garbie et al., 2008a). Agile system does not represent a series of techniques much as it represents a fundamental change in production and/or management philosophies (Gunasekaran et al., 2002). These improvements required are required not only in a small scale but in a completely different way of doing business with the primary focus of flexibility and quick response to changing markets as well.

As agility is used to update the level of manufacturing firms for competition or industry modernization programs, this new concept *non-traditional* or *nonconventional*, or *nonclassic* should be introduced into manufacturing firms to assess the competitive strategy of these firms. Evaluations of manufacturing firms non-traditionally are still the most important issue for the next period, and it will be highly considered. This will lead to a great change in the traditional manufacturing organizational/firms. There will be changes in production such that manufacturing firms will quickly respond to customer demand with high quality in compressed time. On the other side, it can be found that the traditional manufacturing workers on the shop floor will focus on their own small portion of the process without regard to the next step. There will be other changes in some areas such as the following: production support, production planning and control, quality assurance, purchasing, maintenance, marketing, engineering, human resources, finance, and accounting. These changes will cause a revolution in the manufacturing enterprises (Garbie et al., 2008a).

However, there is a need for a systemic approach to evaluate and study the nonconventional performance measurements in manufacturing enterprises. Therefore, there is a strong relationship between lean production and agile manufacturing. Measuring leanness and agility must be related and integrated based on the complexity of the system.

2.3 Analysis of Non-Traditional Evaluation Aspects

Non-traditional aspects (complexity, agility, and leanness) are still ambiguous and an ill-structured problem because they are subjectively described assessments and are unsuitable and ineffective classical techniques. Regarding complexity, there are four important questions to be asked concerning manufacturing complexity (Garbie and Shikdar, 2011a) as follows:

- How is the complexity level of a firm estimated?
- How can a firm reduce its complexity?
- Which issues are more important than others?
- How can firms identify the adverse factors for reducing complexity?

Regarding leanness aspect, there are also some comments to be discussed before analyzing the manufacturing leanness such as the following (Garbie, 2010):

- Value-added and non-value-added activities.
- Which lean manufacturing techniques can be used?
- How the non-value-added can be eliminated?

With respect to manufacturing agility, there are six important questions to be asked concerning agility as follows (Garbie et al., 2008a):

- How far down the path is a company toward becoming a manufacturing organization?
- How and to what degree does the organizational attributes affect the company's business performance?
- How do you measure or evaluate the agility of a company?
- How can a company improve its agility?
- Which factors are more important than others?
- How can companies identify the adverse factors for improving?

Based on these concepts of *complexity, agility, and leanness*, this proposed evaluation suggests three frameworks to focus on complexity, leanness, and agility, respectively. Regarding *complexity*, the system vision complexity, system structure (design) complexity, system operating, and system evaluation complexity are used as the infrastructures of complexity (see Table 2.1). Also, three major infrastructures of *leanness* (supplier related, customer related, and internally related) are used with their sub-major infrastructures to estimate the manufacturing leanness (see Table 2.2). With respect to *agility*, four infrastructures are used to focus on agile capabilities (technology, people, manufacturing strategy, and management) (see Table 2.3). They are considered to be the pillars of nonconventional performance evaluation of manufacturing enterprises (see Figure 2.1). As the overall problem of performance measurement is limited to the three frameworks, the

Table 2.1 Complexity Aspect and Its Components

ASPECT	THEME	SUBTHEME
Complexity (CL)	System vision (SV)	Time to market
		Supply chain management
		Demand variability
		Introducing no. of new products
		Product life cycle
	System design (SD)	Product structure and design
		System design
		Manufacturing philosophies
	System operating (SO)	Status of operating resources
		Shop floor control
		Work in progress
		Business operations
	System evaluation (SE)	Product cost
		Quality
		Productivity
		Response
		Performance appraisal

Table 2.2 Leanness Aspect and Its Components

ASPECT	THEME	SUBTHEME
Leanness (LM)	Supplier (SC)	Supplier feedback
		Just-in-time (JIT) delivery
		Supplier
	Customer (CU)	Customer
	Internal (IN)	Pull
		Continuous flow
		Setup time reduction
		Statistical process control
		Employees involvement
		Total productive maintenance

major fundamental questions, what to measure, how to measure, and how to evaluate the results, will be determined. The analysis could be performed in an interview survey by quantifying the importance from 1 to 10 based on three concepts of evaluation: optimistic, most likely, and pessimistic. This analysis is also proposed from a manufacturing system analyst's perspective, which means it has some delimitation by distributing a questionnaire among industry experts. These questions might not be enough, but they give an idea of how the company is struggling today and an indication of influences in the future.

Table 2.3 Agility Aspect and Its Components

ASPECT	THEME	SUBTHEME
Agility (AL)	People (PE)	Knowledge and skills of workers
		Workforce empowerment
		Interpersonal skills
		Team-based work
		Job enrichment
		Job enlargement
		Improved workforce capability and flexibility
	Manufacturing strategies (MS)	Virtual manufacturing environment
		Supply chain management
		Concurrent engineering
		Reconfiguration
	Technology (TE)	Production design infrastructure
		Components infrastructure
		Information infrastructure
	Organization management (OM)	Customer oriented
		Time to market for launching new product
		Number of new products produced by factory
		Interdepartmental conflicts

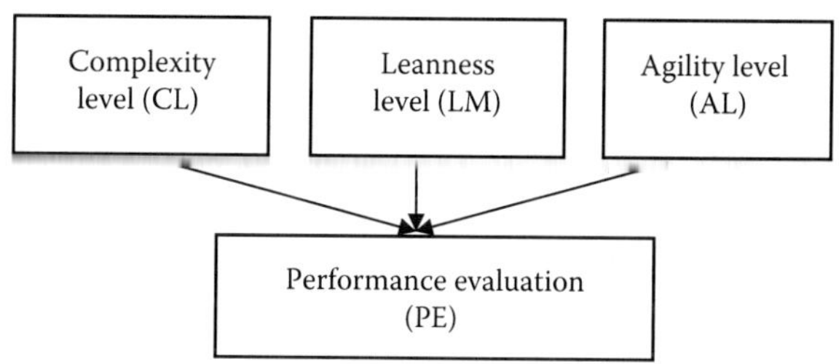

Figure 2.1 Aspects of non-traditional performance evaluation.

2.4 Proposed Methodology

2.4.1 Fuzzy Logic Approach

The basic architecture of each aspect (complexity, agility, and leanness) is depicted in Figure 2.2. In order to perform the aspect evaluation, the system architecture consists of three main parts: fuzzification interface, fuzzy measure, and defuzzification interface. The details of fuzzy logic approach will be discussed in depth through methodology procedure.

2.4.1.1 Fuzzification Interface The variables of basic-level attributes may be expressed with fuzzy values to characterize their uncertainty.

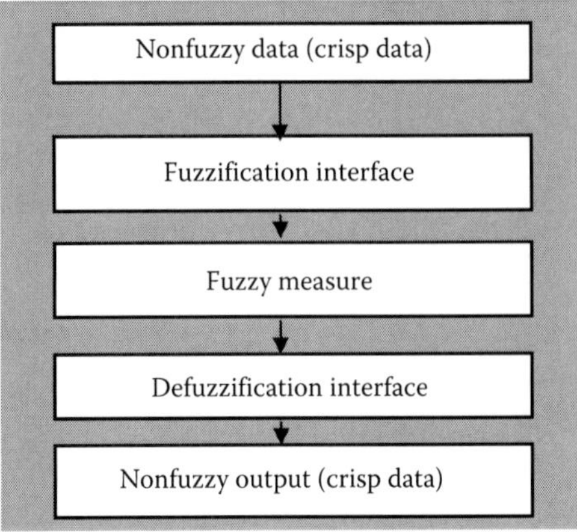

Figure 2.2 Architecture for fuzzy logic approach.

Triangular membership functions were used in this study to express these basic-level attributes. Because the units and the range of raw values for the basic attributes are different, it is difficult to compare them directly. The raw value of each basic variable should be transformed into an index that is bounded in the uniform range from 1 to 10 by using the best value and worst value for the basic attributes. The transformation process normalizes the attribute values in relation to the best and worst values for a particular criterion. The expert assigns the best value (BV) and the worst value (WV) for a particular attribute. The linear transformation index value $\mu(x_i)$ can be calculated for the raw value of each attribute, Z_i, as follows (Garbie et al., 2008a; Garbie and Shikdar, 2011a):

$$\mu(x_i) = \frac{Z_i - \text{WV}}{\text{BV} - \text{WV}} \tag{2.1}$$

where

Z_i is the raw value of each attribute or each question ($\text{WV} < Z_i < \text{BV}$)
$\mu(x_i)$ is the linear transformation index value (membership)
BV is the best value = 10
WV is the worst value = 1

2.4.1.2 Fuzzy Measure Measurement of the fuzziness $f(\hat{A})_j$ of an infrastructure $\hat{A}$ individually is as follows (Garbie et al., 2008a; Garbie and Shikdar, 2011a):

$$f(\hat{A})_j = 1 - \left[\sum_{i=1}^{n(\hat{A})} \left| \mu_{\hat{A}}(x_i) - \mu_{\not\subset \hat{A}}(x_i) \right|^p / (n_{(\hat{A})}) \right]^{1/p} \tag{2.2}$$

where

p = 2 at the Euclidean metric

$n_{(\hat{A})}$ is the number of attributes (questions) in each infrastructure

$\mu_{\hat{A}}(x_i)$ is the membership function

$\mu_{\not\subset \hat{A}}(x_i) = 1 - \mu_{\hat{A}}(x_i)$, j = status of fuzzy member triangle (pessimistic, optimistic, and most likely)

Each status was given a relative score, and the measuring of fuzziness $f(\bar{D})_j$ of each aspect (D) is estimated as follows:

$$f(D)_j = 1 - \frac{\left[\sum_{i=1}^{n_{infrastr}} \left(2\mu_{infra}(x_i) - 1 \right)^2 \right]^{1/P}}{\left\| n_{infra} \right\|^{1/P}} \tag{2.3}$$

2.4.1.3 Defuzzification Interface The defuzzification can be calculated as follows (Garbie et al., 2008a; Garbie and Shikdar, 2011a):

$$\bar{X} = \frac{p + 2m + o}{4} \tag{2.4}$$

where

p is pessimistic

o is optimistic

m is most likely

The output domain $\bar{X}$ is a unique solution and uses all the information of the output membership function distribution.

2.4.2 Methodology Procedure

The proposed methodology will be adapted to combine all dimensions and their corresponding infrastructures to determine the overall performance of complexity, agility, and leanness (see Figure 2.2). All these issues will be explained in the following steps:

Step 1: Questionnaires are designed for each issue including all essential elements regarding complexity, agility, and leanness.

Step 2: Questionnaires are distributed to specific experts in different departments.

Step 3: Questionnaires containing raw values are gathered separately.

Step 4: Raw data are aggregated.

Step 5: Data that are coming from questionnaire are divided into the infrastructures.

Step 6: The fuzzification interface for dimension infrastructures is used to transform crisp data into fuzzy data using Equation 2.1.

Step 7: The measure of the fuzziness (f) of each aspect is used as in Equation 2.2.

Step 8: The aggregate measure (agg.) of the fuzziness (f) for all infrastructures regarding each aspect is determined using Equation 2.3.

Step 9: Evaluate the defuzzification values using Equation 2.4. The output from *Step 8* is a fuzzy membership function for the manufacturing firm's dimension level (complexity, agility, and leanness), which can be defuzzified to yield a nonfuzzy output value (crisp data are needed) from an inferred fuzzy output.

Step 10: Assess the current manufacturing firm's dimension level. The output from *Step 9* is the current value of the manufacturing firm's dimension level.

Regarding the evaluation of performance measurement, the expected value of non-traditional performance measures (PE) is clearly expressed as follows:

$$\text{PE} = f(\text{CL}, \text{AL}, \text{LM}) \quad (2.5)$$

Equation 2.5 can be rewritten with different nomenclatures as follows:

$$\text{PE} = w_{CL}(\text{CL}) + w_{AL}(\text{AL}) + w_{LM}(\text{LM}) \quad (2.6)$$

where

CL is the complexity level
AL is the agility level
LM is the leanness level

The symbols w_{CL}, w_{AL}, and w_{LM} are the relative weights of complexity, agility, and leanness levels, respectively. The value of these weights may reflect the system designer's subjective preferences based on his/her experience or can be estimated using tools such as analytical hierarchy process (Garbie et al., 2008b).

2.5 Case Study and Implementation

During our research, the proposed approach was conducted in one manufacturing firm (XYZ Company) for validation. The objective of this study was to analyze and measure complexity, agility, leanness, and the associated performance measurement. Each dimension was evaluated based on interview surveys and questionnaire that were distributed to their concerned members. The level of complexity was estimated according to 4 infrastructures; also the agility was evaluated based on 4 infrastructures; and leanness was evaluated based on 10 infrastructures. Brief description of this company along with their score for complexity, agility, and leanness is given later. The XYZ Company is the only Aluminum extrusion company in Oman. The company has a unique process of powder coating applications for the manufacturing of aluminum products and accessories. More than 80% of its products are used by construction and architectural sector for doors and windows. The product range includes mill finish products, powder spray painting, anodizing, and wood-coated products. The production plant is operating at full capacity in three shifts and has a capacity to produce over 1800 metric tons per annum of high yield strength aluminum bars of different sizes. The plant is equipped with highly automated two extrusion machines based on German and Italian technologies. One of them was installed in 1985, and the second was installed later in 2003. The company purchases aluminum billets from a major supplier in Dubai (UAE). The material is heated in furnace from 400°C to 520°C and then cut into pieces to get the required shape and dimensions according to the specification of customers. The hot finished bars are subjected to an online heat treatment process to improve mechanical properties. To accomplish design and specification needs of the customer, the dies are kept in stock that can be retrieved easily through automated computerized storage and retrieval system; different colors and codes are also used for material identification purposes. This advanced material storage system facilitates company to produce aluminum bars with

Table 2.4 Results of XYZ Company

DIMENSION/ASPECT	EVALUATION (INDIVIDUAL)	RELATIVE WEIGHTS (%)
Complexity (CL)	0.4460	38
Leanness (LM)	0.2394	29
Agility (AL)	0.4640	33
Performance evaluation (PE)	0.40	

great variety. Online monitoring, offline quality checks, and preventive maintenance practices are applied to achieve precision and high level of quality. The company has ISO certification (e.g., 9000–9001), which ensures prime quality of their products. Almost 80% material is recovered from raw material while the remaining 20% scrap is send back to original supplier. The company is a major exporter and supplied 60% of its product to Gulf countries; 12% to Europe; almost 20% to other neighboring countries like India, Pakistan, and Africa; and the remaining 8% to Omani society. Table 2.4 shows the level of complexity, agility, leanness, and the associated performance evaluation of this company.

It can be noticed from Table 2.4 that the complexity level was estimated at 0.4660 measured on the scale 0–1; leanness level and agility level were estimated at 0.2394 and 0.4640, respectively, on the same scale. The associated performance evaluation was 0.40 measured on the same scale (0–1). The results show that levels of complexity and leanness are medium and low, respectively, and this will be fine, and more reduction in both levels is still needed especially in complexity. Regarding agility, the level of agility is considered as medium although level of leanness is low. More increases in the agility level are required. The performance evaluation of the manufacturing firm is also considered as medium.

2.6 Conclusions

Estimation of the manufacturing complexity level, manufacturing agility level, and manufacturing leanness is a very critical issue of manufacturing firms to survive in the global competition. They are considered the non-traditional performance indicators of manufacturing firms. Analysis and measurements of manufacturing firms for these non-traditional indicators were proposed and presented deeply for each one individually. As a consequence of this, the fuzzy logic approach was proposed to measure the complexity level of the whole firm. With respect to complexity, it was measured through three main infrastructures. Regarding manufacturing agility, it was introduced through four infrastructures. For manufacturing leanness, it was presented through three main infrastructures. Evaluating manufacturing organizations by three subjective status (optimistic, most likely, and pessimistic) will add credibility and objectivity to the evaluation. The results show acceptable variation in these non-traditional performance measurements compared to what was seen in reality in particular organizations. It can be observed from this manufacturing firm that collecting and analyzing the huge amount of data was challenging and time consuming, and it is not an easy task. Application of the proposed non-traditional manufacturing performance evaluation in a real case study is presented and illustrated to explain which areas are needed to be analyzed and studied.

References

Bozarth, C.C., Warsing, D.P., Flynm, B.B., and Flynn, E.J. 2009. The impact of supply chain complexity on manufacturing plant performance. *Journal of Operations Management*, 27, 78–93.

Cho, S., Alamoudi, R., and Asfour, S. 2009. Interaction based complexity measures of manufacturing systems using entropy. *International Journal of Computer Integrated Manufacturing*, 22(10), 909–922.

Garbie, I.H. 2009. A vision for reconfiguring industrial organization due to the global recession. *Proceedings of the 39th of International Conference on Computers and Industrial Engineering*, Troyes, France, pp. 658–663, July 6–8, 2009.

Garbie, I.H. 2010. A roadmap for reconfiguring industrial enterprises as a consequence of global economic crisis (GEC). *Journal of Service Science and Management*, 3(4), 419–428.

Garbie, I.H. 2012a. Design for complexity: A global perspective through industrial enterprises analyst and designer. *International Journal of Industrial and Systems Engineering*, 11(3), 279–307.

Garbie, I.H. 2012b. Concepts and measurements of industrial complexity: A state of the art survey. *International Journal of Industrial and Systems Engineering*, 12(1), 42–83.

Garbie, I.H. 2013a. DFMER: Design for manufacturing enterprises reconfiguration considering globalization issues. *International Journal of Industrial and System Engineering*, 14(4), 484–516.

Garbie, I.H. 2013b. DFSME: Design for sustainable manufacturing enterprises (an economic viewpoint). *International Journal of Production Research*, 51(2), 479–503.

Garbie, I.H., Parsaei, H.R., and Leep, H.R. 2008a. A novel approach for measuring agility in manufacturing firms. *International Journal of Computer Applications in Technology*, 32(2), 95–103.

Garbie, I.H., Parsaei, H.R., and Leep, H.R. 2008b. Measurement of needed reconfiguration level for manufacturing firms. *International Journal of Agile Systems and Management*, 3(1/2), 78–92.

Garbie, I.H. and Shikdar, A. 2011a. Analysis and estimation of complexity level in industrial firms. *International Journal of Industrial and Systems Engineering*, 8(2), 175–197.

Garbie, I.H. and Shikdar, A. 2011b. Complexity analysis of industrial organizations based on a perspective of systems engineering analysis. *Journal of Engineering Research* (*TJER*), SQU, 8(2), 1–9.

Giachetti, R.E., Martinez, L.D., Saenz, O.A., and Chen, C.-S. 2003. Analysis of the structural measures of flexibility and agility using a measurement theoretical framework. *International Journal of Production Economics*, 86, 47–62.

Groover, M.P. 2001. *Automation, Production Systems, and Computer-Integrated Manufacturing*, 2nd edn., Prentice Hall, Upper Saddle River, NJ.

Gunasekaran, A., Tirtiroglu, E., and Wolstencroft, V. 2002. An investigation into the application of agile manufacturing in an aerospace company. *Technovation*, 22, 405–415.

Hu, G., Wang, L., Fetch, S., and Bidanda, B. 2008a. A multi-objective model for project portfolio selection to implement lean and six sigma concepts. *International Journal of Production Research*, 44(23), 6611–6625.

Hu, S.J., Zhu, X., and Koren, W.Y. 2008b. Product variety and manufacturing complexity in assembly systems and supply chains. *Annals of the CIRP, Manufacturing Technology*, 57, 45–48.

Jenab, K. and Liu, D. 2010. A graph-based model for manufacturing complexity. *International Journal of Production Research*, 48(11), 3383–3392.

Kuzgunkaya, O. and ElMaraghy, H.A. 2006. Assessing the structural complexity of manufacturing systems configurations. *International Journal of Flexible Manufacturing Systems*, 18, 145–171.

Tani, G. and Cimatti, B. 2008. Technological complexity: a support to management decisions for product engineering and manufacturing. *2008 IIIE International Conference on Industrial Engineering and Engineering Management (IEEM 2008)*, Singapore, pp. 6–11.

Thomas, A., Franai, M., John, E., and Davies, A. 2012. Identifying the characteristics for achieving sustainable manufacturing companies. *Journal of Manufacturing Technology Management*, 23(4), 426–440.

Wu, Y., Frizelle, G., and Efstathiou, J. 2007. A study on the cost of operational complexity in customer-supplier systems. *International Journal of Production Economics*, 106, 217–229.

Yang, J. 2010. A new complexity proof for the two-stage hybrid flow shop scheduling problem with dedicated machines. *International Journal of Production Research*, 48(5), 1531–1538.

3

Automotive Stamping Operations Scheduling Using Mathematical Programming and Constraint Programming

BURCU CAGLAR GENCOSMAN, H. CENK OZMUTLU, HUSEYIN OZKAN, AND MEHMET A. BEGEN

Contents

3.1 Introduction

The stamping operations scheduling problem is one of the significant real-world scheduling problems that have not been investigated in literature in detail so far, although scheduling of stamping operations is a crucial process for automotive stamping companies. Beycelik Gestamp is one of the leader companies in the field of sheet metal forming and die production in Bursa, Turkey, which works for main automotive companies such as Fiat, Ford, Renault, Volkswagen, and

Maserati. The company produces external parts like sides and roofs, and internal parts like bumpers of an automotive. These parts are produced in different stamping *presslines* that are grouped as robotic presslines and manual presslines. The robotic presslines work automatically without operators and includes a small number of presses in sequential. On the other hand, the manual pressline needs operators for loading and unloading the products. The stamping scheduling problem involves assignment and sequencing of metal sheets on these presses and scheduling them in such a way that minimizes production time and satisfies all the demand. Products require different number of operations (presses), and presses are not identical. The details of the problem that we consider in this chapter are elaborated in Section 3.2. Next, we present a brief literature review.

Many researchers developed different solution methods for different real-world scheduling problems (Khayat et al. 2006, Barlatt et al. 2012, Relvas et al. 2013). Barlatt et al. (2012) investigated the stamping operations of Ford Motor Company. They took into account a stamping pressline with identical presses that include jobs with one operation. Therefore, they converted the scheduling problem to a task sequencing problem and developed a Test-and-Prune algorithm. In addition, they considered the shift selection problem with task sequencing problem, and they developed a decision support tool: the just-in-time execution and distribution information system (JEDI). The researchers monitored the production environment completely, including supply chain and workforce allocation by the JEDI system. They also developed a stamping scheduling optimizer (SSO) by composite decision variables for the production planning, scheduling, and the workforce allocation. They first generated the initial schedules by JEDI and sent them to SSO rather than using mixed-integer programming (MIP) approach. The researchers were able to generate effective scheduling plans by their method. Unlike this study, our problem includes scheduling of jobs with different amounts of operations on nonidentical presses. These jobs can be assigned simultaneously while the total number of operations is smaller or equal to the total number of presses. Therefore, our problem becomes unrelated parallel machine scheduling problem (RCPMSP).

The parallel machine scheduling problems (PMSPs) are one of the most widely studied problems in literature (Edis et al. 2013).

Edis et al. (2013) reviewed the PMSPs with additional resources. They evaluated the PMS studies with respect to five main topics. They also demonstrated the complexity of the resource-constrained scheduling problems considering the study of Błażewicz et al. (2007). Edis and Ozkarahan (2012) researched a real-life resource-constrained PMSP in electrical appliance plant. They generated a solution method by integer programming (IP), but they could not reach the optimality because of the huge number of variables and constraints. Therefore, the researchers tried to simplify the main problem by partitioning it into two subproblems. They also used constraint programming (CP) with IP to develop solution methods for the subproblems. They solved the subproblems by combining IP/IP and IP/CP. Although the IP/IP model reached better results in test problems, the IP/CP model generated reasonable solutions in less than 1 minute when the resource constraints were tight.

CP applications become popular in the field of operations research for generating the exact solutions of scheduling problems (Van Hentenryck 1999). Malapert et al. (2012) developed a CP approach for a batch processing problem with nonidentical job sizes. The researchers also used a mathematical programming model to compare the experimental results with CP. Experimental results demonstrated that the CP outperforms the mathematical programming in terms of computation time and quality gap of a solution. Consequently, the CP approach may be an alternative solution method to mathematical programming.

In this chapter, we first present a mixed-integer program (MIP1) that models the scheduling problem. We also use MIP1 to test the company's current practice of 4 hour production period length. However, MIP1 is not effective in solving for the real-world instances of the problem. Therefore, we develop a constraint programming (CP1) model and compare its results with results of the MIP1 model. We also develop/establish improvement studies on our solution methodologies.

The chapter is organized as follows. In Section 3.2, we give a formal problem description of the stamping operations scheduling problem in Beycelik Gestamp. In Section 3.3, we define the MIP1 model with its use. In Section 3.4, we present the CP1 model. Improvement studies are given in Section 3.5. In Section 3.6, we compare our proposed solutions based on the models with real-world schedules. We conclude the chapter in Section 3.7.

3.2 Stamping Operations Scheduling Problem

We consider the largest manual pressline that includes 13 sequential presses. Different from the studies in literature, each product requires different number of operations to become a final product, which must be assigned to consequent presses. In addition, each operation of a product needs an exclusive upper and lower *die pair* that is loaded to related press at the beginning of the production. The die pair forms the raw material, a steel *blank*, while it passes through the pressline. The blank becomes a final product after completing its operations. The company can produce different kinds of products simultaneously given that the total number of operations is less than or equal to the total number of presses. For example, four distinct products with 2, 3, 4, and 4 operations with different durations can be produced in the pressline at the same time as seen in Figure 3.1. The first operation of Product C (C.1) starts at Press 6 (P6), and the blank moves through the pressline until Press 9 (P9). At the end of P9, the blank of Product C can be categorized as the final product.

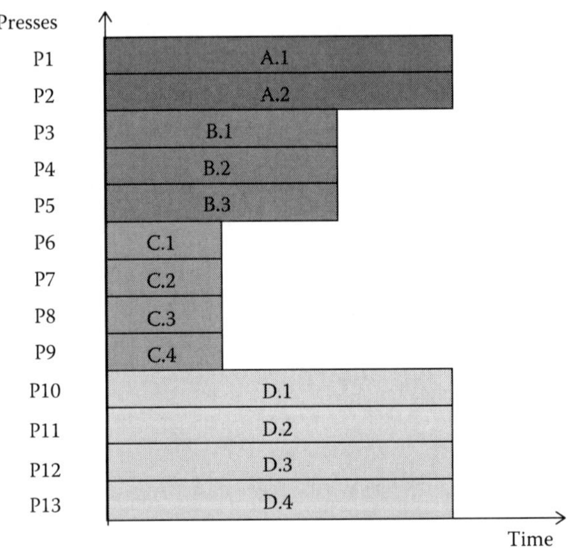

Figure 3.1 Loading of four different products to the pressline.

Considering the review of PMS problems (Edis et al. 2013), we define our specific stamping scheduling problem. Edis et al. (2013) define RCPMSP as a set of n jobs to be processed on a set of m unrelated parallel machines. Also, they mention that if machine-related resources (e.g., dies) are limited, they should be evaluated as resource constraints. In our stamping problem, we have a set of n products (jobs) with different number of j operations, which must be processed at j consecutive presses (machines) in a set of m unrelated parallel machines. In addition, the machines are not identical, so jobs must be assigned according to machine eligibility restrictions. Each operation of a job needs a die pair that is limited. Thus, we classify our stamping scheduling problem as an RCPMSP with multiple consecutive operations and machine eligibility restrictions. To the best of our knowledge, this problem has not been studied in the literature.

The stamping scheduling in Beycelik poses several difficulties. The first and the most important difficulty is the fluctuation of customer demand during the production horizon. Although the company works with weekly demands, customers tend to change their order amount or place urgent orders. The company tries to respond to these implementations as quickly as possible by changing their weekly schedule.

The second difficulty is the breakdown of machines. When a breakdown occurs, the operation of the related job must be postponed until a new machine is found or the breakdown is repaired. Where the breakdown occurs adversely affects the other operations of the job. If the breakdown occurs at the beginning machines of operations instead of the last one, the other operations of the job must be postponed. If the breakdown occurs at the last machine of the operations, the related die pair is moved to another machine if available. In this case, the first operations of the job continue to process and then semiproducts are moved to new machine in order to complete the job. For example, in Figure 3.1, if the breakdown happens at P7, the other operations of Product C must stop until P7 is repaired, which causes the stoppage of the other three machines. But if the breakdown occurs at P9, the die pairs are loaded to another available machine, and the semiproducts that are produced at the first three machines are moved to new machine to complete the production of Product C. However, machines are not uniform in terms of weight, pressure, and size; hence, finding an available machine is difficult. For example, if an operation requires high pressure, the related dies cannot be loaded to low-pressure machines. On the other hand, if an operation requires a low-pressure machine, the related dies can be loaded to high- and low-pressure machines. As a result of this variability, called *machine eligibility restrictions* (Pinedo 2012), the operations of the job must be assigned considering the property of the machine and the die pair; thus, we have to define *alternative machine sets* for each job. Therefore, jobs must be scheduled according to their alternative machine sets, and this unavoidable situation makes generating effective schedules difficult.

The third difficulty is to determine the effective changeovers from one job to another. The changeover of dies takes 30 minutes on average and decreases the total production time. During the changeovers, production must stop and the related machines must be idle, which result in loss for the company. The company plans and implements that the changeover from one job type to another is completed during the lunch break or during the shift changes. This means a 4 hour production period, that is, when a job is assigned to a related machine, it must be produced for at least 4 hours. Although this 4 hour production period practice makes changeovers independent from the total production time, it sometimes forces production to exceed demand. If a job cannot be completed in 4 hours, the related dies will stay at the same machines, and the production of this job will be continued (at least) for the next 4 hours (the next period). On the other hand, if a job is completed in less than 4 hours, the related dies will still remain in the same machines until to the end of period, which may cause production to exceed demand.

To evaluate the effectiveness of the 4 hour production period practice, we use MIP1 to find an optimum period length considering the trade-off between production amount and changeovers. The company assumes that jobs may not be preempted, and demand must be met by one-time loading of die pairs. However, investigating the trade-off between production amount and changeovers requires separation of jobs. Therefore, we use MIP1 with the assumption that jobs may be preempted considering the changeovers in Section 3.2. We next define the MIP1 model according to the conventions and practices of company and the problem constraints.

3.3 Mixed-Integer Programming Model: MIP1

The Beycelik Gestamp assumes that jobs may not be preempted, and once the production of a job is started, it must be produced to satisfy at least its demand. In order to model and reflect the real system accurately, we develop the MIP1 model with the assumption that jobs may not be preempted and with the following constraints:

- Every job has a sequential number of operations and these must be assigned to sequential machines.
- Every job must be assigned to only one machine during one period.
- Every machine must produce only one job during one period.
- The total amount of production must be equal to or greater than the demand.
- The changeover occurs at the beginning of a period and/or at the end of a period.
- The period length is 4 hours; hence, if job i is starts at machine m, it must remain for at least 4 hours.

We next provide the details of the parameters and constraints of MIP1.

Indices (for sets, see below)

i: Index of jobs to be scheduled, $i \in I$.
k: Index of periods, $k \in K$.
m: Index of machines, $m \in M$.
j_i: Index of operation numbers, $j \in O_i$.

Parameters

R: Changeover time of dies for each job (30 minutes).
D_i: Demand of job i.
A_{im}: 0–1 matrix of alternative machines of jobs. A_{im} is 1 if job i can be assigned to machine m and zero otherwise.
P_{ij}: Processing time of operation j for job i.
$P_{i\max}$: denotes the longest processing time of all operations for job i
J_i: Total duration of job i.
U: Period limit in hours.
L_1: Large number (at least maximum demand + 1).

L_2: Large number (at least maximum number of machines + 1, which is fixed to 14 in this problem).
L_3: A positive number to allow and to prevent the preemption of jobs.

Sets ($I_{\max}$, $K_{\max}$, $M_{\max}$, and $O_{i\max}$ denote the largest index value of the sets I, K, M, and O_i, respectively.)

I: The set of jobs. $I=\{1, 2, \ldots, I_{\max}\}$.
K: The set of periods. $K=\{1, 2, \ldots, K_{\max}\}$.
M: The set of machines. $M=\{1, 2, \ldots, M_{\max}\}$.
O_i: The set of operations for job i. $O_i=\{1, 2, \ldots, O_{i\max}\}$.
$Set\,1 = \{<i, k, m> \mid i \in I, k \in K, m \in M, A_{im} = 1\}$.

The required number of periods adversely affects the dimension of the problem. Therefore, to decrease the solution space, we determine the required number of periods ($K_{\max}$) according to the first feasible point of the MIP1 solution. For example, if $K_{\max}$ is equal to 40 while the problem is infeasible, and we reach the first feasible solution with $K_{\max}$ is equal to 42, and then we define the number of periods as 42 for the related problem.

Decision variables

$$x_{ikm} : \begin{cases} 1, & \text{if job } i \text{ is processed in period } k \text{ at machine } m \\ 0, & \text{otherwise} \end{cases}$$

a_{ik}: Amount of job i produced in period k.
c_{ikm}: Changeover time for job i in period k on machine m.
$C_{\max}$: Makespan

We next present the model:

$$\min \ z = C_{\max} + L_3 \times \sum_i \sum_k \sum_m c_{ikm} \tag{3.1}$$

$$(k-1) \times U \times x_{ikm} + \frac{\sum_{j \in O_i} P_{ij} + (a_{ik} - 1) \times P_{i\max}}{3600} + \sum_{\substack{<e,l,n> \in Set\,1 | 0 \le l \le k+1, \\ n \le m,\, n+O_e-1 \ge m}} c_{eln} \le C_{\max}, \quad \forall (i,k,m) \in Set\,1 \tag{3.2}$$

$$\sum_{j \in O_i} P_{ij} + (a_{ik} - 1) \times P_{i\max} \leq 3600 \times U, \quad \forall i \in I, k \in K \tag{3.3}$$

$$\sum_{i \in I | <i,k,m> \in Set\ 1} x_{ikm} \leq 1, \quad \forall k \in K, m \in M \tag{3.4}$$

$$\sum_{m \in M | <i,k,m> \in Set\ 1} x_{ikm} \leq 1, \quad \forall i \in I, k \in K \tag{3.5}$$

$$a_{ik} \leq L_1 \times \sum_{m \in M | <i,k,m> \in Set\ 1} x_{ikm}, \quad \forall i \in I, \ k \in K \tag{3.6}$$

$$\sum_{k \in K} a_{ik} \geq D_i, \quad \forall i \in I \tag{3.7}$$

$$x_{tkn} \leq 1 - x_{ikm}, \quad \forall (i,k,m) \in Set\ 1, (t,k,n) \in Set\ 1 \,|\, t \neq i, \quad m \leq n \leq m + O_{i\max} - 1 \tag{3.8}$$

$$x_{ikm} \times m + O_{i\max} - 1 - L_2 \times (1 - x_{ikm}) \leq M_{\max}, \quad \forall (i,k,m) \in Set\ 1 \tag{3.9}$$

$$R \times (x_{ikm} - x_{i,k+1,m}) \leq c_{ikm}, \quad \forall (i,k,m) \in Set\ 1 \,|\, k+1 \leq K \tag{3.10}$$

$$R \times (-x_{ikm} + x_{i,k+1,m}) \leq c_{ikm}, \quad \forall (i,k,m) \in Set\ 1 \,|\, k+1 \leq K \tag{3.11}$$

$$a_{ik}, c_{ikm} \geq 0, \quad \forall i \in I, k \in K, m \in M \tag{3.12}$$

$$x_{ikm} \in \{0,1\}, \quad \forall (i,k,m) \in Set\ 1 \tag{3.13}$$

The aim of MIP1 is to minimize the total duration of production horizon, the makespan ($C_{\max}$) as seen in expression (3.1). We can allow the preemption of jobs by assigning zero to L_3, and we can prevent the preemption of jobs by assigning a large positive number to L_3 so that we can compare MIP1 with CP1 logically and represent the real system accurately. Constraint (3.2) represents the calculation of makespan, which also includes changeovers. The maximum completion time of each job can be calculated by multiplying the previous periods with period limit, adding the production time at the last period, which is switched to hours by dividing 3600 seconds, and considering the changeover times of previous operations. The first product is completed after all operations are finished, but it is not the same for the rest of the production amount. After the first operation is completed for the first blank, the blank moves to the second machine for the second operation and the second blank is loaded to the first machine. Therefore, the total completion time of the first job is the total processing time of its operations, but for the rest of the production amount ($a_{ik} - 1$), the total completion time would be the longest operation's processing time. It has to be mentioned that, in our problem, the processing time of operations of a job is equal to each other. Moreover, the processing time in period k for job i must be limited by the period limit as seen in constraint (3.3). Constraint (3.4) restricts the production of jobs in different machines during the same period. Constraint (3.5) indicates that one machine must produce only one job during one period. Constraint (3.6) ensures that if job i is not assigned to machine m at period k, the amount of production of this job at period k would be zero, and constraint (3.7) indicates that the total production amount of a job must be equal to or greater than its demand. Constraint (3.8) guarantees that if job i is assigned to machine m at period k, no other jobs can be assigned until its required number of operations are completed. Constraint (3.9) ensures that if job i is assigned to machine m at period k, the total amount of its starting point and the number of operations must be smaller than or equal to the maximum number of machines in the pressline. Constraints (3.10) and (3.11) calculate the changeover time depending on the position of a job. Constraints (3.12) and (3.13) define the continuous and binary decision variables. To simplify the model, we assume that the amount of production i (a_{ik}) is a continuous decision variable rather than integer. (This is justified as the number of items produced is large.)

To evaluate the performance of MIP1, we use real production data. In general, the company schedules at least 50 different jobs in a week, and as seen in Table 3.1, we have 52 different jobs to schedule. We generate 10 different problems from real production data, which have different number of jobs by increasing the number of jobs for each problem iteratively. The system is modeled by ILOG CPLEX 12.4, and its solution time is limited to 1800 seconds. We run the experiments on a PC with 2.90 GHz i7 processor and 6 GB RAM and detailed the results in Table 3.1. The first three columns present information about the instances such as the number of jobs and the

Table 3.1 MIP1 Performance on Real Scheduling Data

PROBLEM	NO. OF JOBS	NO. OF PERIODS	C_{MAX} (H)	ELAPSED TIME (S)	GAP%
1	5	5	16.3	1.1	—
2	10	8	29.3	6.7	—
3	15	13	44.9	1800	22.3
4	20	22	77.6	1800	63.7
5	25	25	84.9	1800	78.7
6	30	27	98.9	1800	87.8
7	35	32	119.2	1800	91.3
8	40	40	146.8	1800	95.9
9	45	49	—	1800	—
10	52	54	—	1800	—

required number of periods. The fourth column indicates the value of objective function (C_{max}), the fifth column shows the elapsed time, and the final column illustrates the gap of the solution. The gap is calculated by the branch-and-cut algorithm in CPLEX 12.4 considering the best integer solution z^* and the best linear solution z' as follows:

$$\text{Gap} = \frac{|z' - z^*|}{e^{-10} + |z^*|} \tag{3.14}$$

According to the C_{max} column of Table 3.1, MIP1 could not find the optimal schedules of problems after Problem 2. In addition, the gap exceeds 95.9% after Problem 8, which shows that we are too far from the optimal solution. Because of the complex nature of the problem, MIP1 spends too much time to find a feasible solution, and it terminates before finding an effective solution. In addition, MIP1 could not find any feasible solution after Problem 9; hence, we need to investigate a better solution method to gather successful schedules faster than MIP1 and to reach feasible solutions of real-world problems. We use MIP1 to evaluate the convention of 4 hour period length in Section 3.3.1.

3.3.1 Evaluation of the Period Length

To evaluate the accuracy of the period length (4 hours) convention of the company, we modify MIP1 with the assumption that jobs may be preempted by assigning zero to L_3. We use the uniform distributions

Table 3.2 Uniform Distributions for the Model Parameters

PARAMETERS	UNIFORM DISTRIBUTION: $U(A, B)$
The number of operations	$U(1, 6)$
The processing times of operations (s)	$U(5, 20)$
The number of alternative machines	$U(1, 3)$
Sets of alternative machines	$U(1, 13)$

in Table 3.2 to generate different instances with different number of jobs. MIP1 is able to solve at most 10 jobs optimally; hence, we evaluate five different problems that contain 10 jobs and present the results in Table 3.3.

The first column shows the values of period that varies from 1 to 12 hours; the second part represents the elapsed time, which is limited to 1800 seconds. The third part illustrates the makespan of the optimal schedule (C_{max}), and the fourth part indicates the required number of periods. The changeover (R parameter) of dies for one job takes 30 minutes. According to Table 3.3, if we assume a period length of 1 hour, we need 34 periods to schedule the first problem P1, and we find a feasible schedule with 29.4 hours in time limits. If we change the period length to 2 hours, we need 15 periods and C_{max} reduces to 29 hours, and it varies from 29 to 31 hours for different lengths of periods.

Since the required number of periods is related with the decision variable in MIP1, the dimension of this index adversely affects the problem size. As seen in Table 3.3, decreasing the period length increases the required number of periods and the number of decision variables. In this situation, problems cannot be solved optimally within time limits, but C_{max} values are better than the other period lengths. Therefore, we have to define a period length considering the trade-off between C_{max} and problem complexity. The C_{max} variation through period length can be demonstrated as in Figure 3.2.

Although the C_{max} of the problems have fluctuations, they tend to increase after 4 hours. Since the changeover time is equal when different period lengths are used, we do not need to consider this parameter. Moreover, when the period length increases, the scheduling problem becomes easier to solve, and we can find the optimal schedules in shorter times. In summary, we should define a large period length to reduce the problem complexity and a small period length to reduce

Table 3.3 Period Length Experiments with Randomly Generated Instances

PERIOD	ELAPSED TIME (S)					C_{MAX} (H)					REQUIRED NUMBER OF PERIODS				
LENGTH (H)	P1	P2	P3	P4	P5	P1	P2	P3	P4	P5	P1	P2	P3	P4	P5
1	1800	1800	1800	1800	1800	29.4	66.1	45.7	63.4	57.4	34	67	46	64	58
2	1800	1800	1800	1549.3	1141.4	29.0	68.1	45.6	64.6	57.0	15	38	24	37	30
3	1800	1800	1800	71.9	402.2	31.0	69.6	45.6	63.6	60.5	11	25	16	23	22
4	99.5	1800	465	15.4	21.6	30.3	72.0	45.6	64.6	56.8	8	20	12	17	16
5	17.3	932.6	134.1	10.8	7.1	30.3	75.0	45.6	65.1	58.5	7	16	10	14	12
6	29.7	93.2	80.0	5.6	10.5	31.0	73.4	45.6	66.6	60.8	6	13	8	12	11
7	5.0	242.9	23.3	11.5	6.2	31.3	84.4	45.6	70.1	59.5	5	13	7	11	10
8	23.5	92.0	6.5	2.6	6.0	34.3	88.4	45.6	67.1	64.7	5	12	6	9	9
9	38.9	35.7	12.0	2.3	4.9	37.0	82.0	45.6	66.5	66.5	5	10	6	8	8
10	2.5	37.1	2.5	2.7	2.6	31.7	90.0	45.6	72.5	63.5	4	10	5	8	7
11	5.5	11.3	2.7	2.3	2.8	33.7	88.3	45.6	78.5	69.5	4	9	5	8	7
12	1.8	6.5	1.7	1.9	2.9	31.0	87.6	45.6	73.5	72.7	3	8	4	7	7

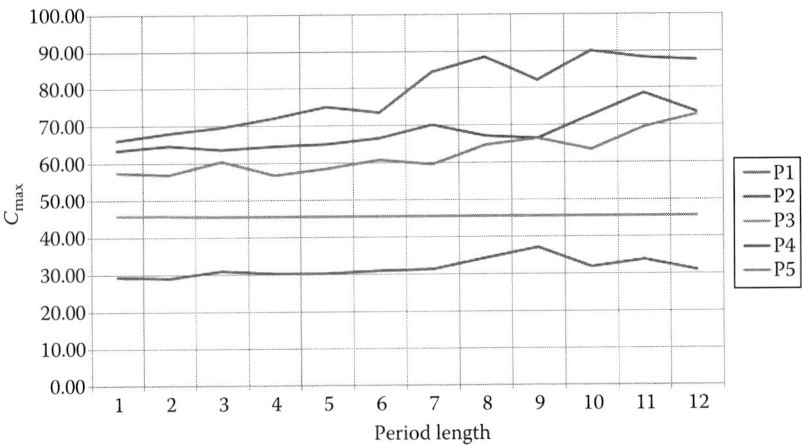

Figure 3.2 The variation of C_{max} within different period lengths.

the unnecessary production of jobs. After considering all these factors and Figure 3.2, we choose the period length of 4 hours, as the company does.

Although MIP1 reflects the real-world stamping scheduling problem accurately, it could not solve instances with 15 jobs or more. Considering weekly demands that contain 50 jobs on average, it is obvious that this model would not be very useful.

On the other hand, CP has become an important tool for generating exact solutions of scheduling problems (Nuijten 1996, Van Hentenryck 1999, Baptiste et al. 2001). It has roots in the Artificial Intelligence community in terms of constraint satisfaction problems (CSPs; Pinedo 2012), and to date, various languages have been developed to combine CP and optimization problems, such as OPL (Van Hentenryck 1999). CP has unique decision variable definitions and simplified constraint expressions (thanks to specialized languages). Utilizing these advantages, we next model the stamping scheduling problem using CP in Section 3.4.

3.4 Constraint Programming Model: CP1

CP has been developed for solving the *CSPs* (Baptiste et al. 2001). The CSP is defined by a set of *variables*, a set of possible values for each variable called *domains*, and a set of *constraints*. The aim of the CSP is to find a possible assignment of values to variables that all constraints are satisfied. The CSP uses the constraints to reduce the solution space by removing the insufficient values from the domains. This reduction is called *constraint propagation*, and it is a feature that distinguishes the CP from mathematical programming. The CP is a CSP with an objective function like a mathematical programming, and we use the *Scheduler* module (developed for scheduling problems) of CP in ILOG CPLEX 12.4 to describe the stamping scheduling problem with its unique expressions.

As discussed before, the Beycelik Gestamp assumes that jobs may not be preempted, and demand must be met in a one-time loading of die pairs. Furthermore, with the help of MIP1, we confirmed that 4 hour period length works well for the company. In order to generate useful real-world schedules with CP1, we use the same conventions with the company; we assume that jobs may not be preempted, and demand must be met by one-time loading of die pairs. The constraints are presented as follows:

- Every job has a sequential number of operations and these must be assigned to sequential machines.
- Every job must be assigned to only one machine during the production horizon.
- Every machine must produce only one job at the same time.
- The total amount of production must be equal to demand.
- The jobs may not be preempted.

Unlike the evaluation strategy of MIP1, CP1 reduces the feasible solution space according to its constraints and domains of variables, and it finds the optimal solution after trying all possible solutions. In addition, the *Scheduler* module has special variable definitions like *intervals*, which can store some information about variables, such as start time, end time, and duration, and special constraint expressions like *alternative*, *span*, and *noOverlap*, which are used to model the stamping scheduling problem. The notation of the CP1 model is presented as follows:

Indices (for sets, see below)

i: Index of jobs to be scheduled, $i \in I$.
k: Index of periods, $k \in K$.
m: Index of machines, $m \in M$.
j_i: Index of operation numbers, $j \in O_i$.

Parameters

D_i: Demand of job i.
A_{im}: 0–1 matrix for alternative machines of jobs. A_{im} is 1 if job i can be assigned to machine m and zero otherwise.
P_{ij}: Processing time of operation j for job i.
J_i: Total duration of job i.
U: Period limit in hours.
H_i: Total periods for job i. Calculated as $H_i = J_i/3600 \times U$.
O_i: Total number of operations of job i.

Sets ($I_{\max}$, $K_{\max}$, $M_{\max}$, and $O_{i\max}$ denote the largest index value of the sets I, K, M, and O_i, respectively.)

I: The set of jobs. $I = \{1, 2, \ldots, I_{\max}\}$.
K: The set of periods. $K = \{1, 2, \ldots, K_{\max}\}$.
M: The set of machines. $M = \{1, 2, \ldots, M_{\max}\}$.
O_i: The set of operations for job i. $O_i = \{1, 2, \ldots, O_{i\max}\}$.

$$Alter = \left\{ < i, m > \mid i \in I, m \in M : A_{im} = 1 \right\}$$

$$Curr = \left\{ < i, j > \mid i \in I, j \in O_i : P_{ij} > 0 \right\}$$

Decision variables

y_i: Interval variable for job i in $[0.3600 \times K_{\max} \times U]$.
e_{ij}: Interval variable for operation j of job i, size J_i/O_i.
x_{im}: Interval variable for alternative machines m of job i in $[0.3600 \times K_{\max} \times U]$.
$g_m = \sum_{<i,j> \in Curr} pulse(e_{ij}, 1), \ \forall m \in M$ (For *pulse*, see below).

In order to attach the related operations to jobs, we define them as *interval* variables. We indicate the definition space of jobs using *in* expression, which means that jobs must be completed between 0 and $3600 \times K_{\max} \times U$ seconds. We define the size of each operation according to the size of job i with $J_i/O_{i\max}$, so the total size of those operations would be equal to the size of job i. As mentioned before, every job has an alternative machine set, and jobs must be assigned according to these sets. Therefore, we define the alternative machines as the *interval* variables and use the *alternative* expression to make sure that every job must be assigned to one of its alternative machines. We assume that every operation causes 1 unit usage of machine, which is added up by *pulse* expression, and we present the *cumulative* usage of each machine with g_m function. According to these variables, the constraints are summarized as follows:

$$\min \max_i(endOf(y_i)) \quad (3.15)$$

$$endBeforeStart(e_{ij}, e_{ij+1}), \quad \forall i \in I,\ j \in O_i \mid j+1 \le O_i \quad (3.16)$$

$$presenceOf(e_{ij}) \xrightarrow{yields} presenceOf(e_{ij+1}), \quad \forall i \in I,\ j \in O_i \mid j+1 \le O_i \quad (3.17)$$

$$span(y_i, all(j \in O_i)e_{ij}), \quad \forall i \in I \quad (3.18)$$

$$alternative(y_i, all(a \ in \ Alter : a.i = i)x_a), \quad \forall i \in I \quad (3.19)$$

$$noOverlap(all(a \ in \ Alter : a.m = m)x_a), \quad \forall m \in M \quad (3.20)$$

$$\begin{pmatrix} presenceOf(x_a) > 0 \\ AND \\ presenceOf(x_b) > 0 \end{pmatrix} \xrightarrow{yields} \begin{pmatrix} endOf(x_a) \le startOf(x_b) \\ OR \\ endOf(x_b) \le startOf(x_a) \end{pmatrix}$$

$$\forall a, b \in Alter \mid O_{a.imax} > 1 \ AND \ b.i \ne a.i \ AND \ b.m \ge a.m \ AND$$
$$b.m \le a.m + O_{a.imax} - 1 \quad (3.21)$$

$$\sum_{m \in M} g_m \le M \quad (3.22)$$

The aim of CP1 is to minimize the maximum completion time of the production horizon. The *endOf* term defines the end time of a job, so expression (3.15) implies the $C_{\max}$. The *endBeforeStart* term prevents e_{ij+1} to start before e_{ij} ends; so constraint (3.16) ensures that the start time of operations must be sequential. The *presenceOf* term is a logical true–false expression, which returns 1–0, and constraint (3.17) links the operations of jobs together. Constraint (3.18) uses *span* expression to make sure that a job must cover its operations in terms of start and end times. Constraint (3.19) guarantees that each job can be assigned to one of its alternative machines. Constraint (20) prevents to assign more than one job to a machine at the same time with *noOverlap*

Table 3.4 Comparison of MIP1 and CP1 with Real Data

			C_{MAX} (H)		ELAPSED TIME (S)		GAP%	ADOL%
PROBLEM	NO. OF JOBS	NO. OF PERIODS	MIP1	CP1	MIP1	CP1	MIP1	CP1
1	5	5	16.3[a]	16.3[a]	1.1	2.9	—	—
2	10	8	29.3[a]	29.3[a]	6.7	2.7	—	—
3	15	13	44.9	**43.9**	1800	109.7	22.3	2.1
4	20	22	77.6	**73.8**	1800	119.5	63.7	4.9
5	25	25	85.0	**78.8**	1800	136.1	78.7	7.3
6	30	27	99.0	**89.0**	1800	141.8	87.8	10.0
7	35	32	119.2	**98.2**	1800	175.2	91.3	17.6
8	40	40	146.8	**126.1**	1800	211.0	95.9	14.1
9	45	49	—	**156.4**	1800	305.7	—	—
10	52	54	—	**170.0**	1800	371.2	—	—

[a] Proven optimality.

Note: Highlighted lower C_{max} values are in bold.

term. Constraint (3.21) ensures that if job a is assigned to machine m, no other jobs can be assigned until its required number of operations are completed. Constraint (3.22) restricts the cumulative usage of machines with machine capacity.

To evaluate the performance of CP1, we used the same 10 instances from real scheduling data as in Section 3.3. We compared CP1 with MIP1, and we restricted MIP1 within 1800 seconds and CP1 with 700,000 fail limits, which is determined in Section 4.1. We provide the results in Table 3.4 and highlighted lower C_{max} values in bold.

The first three columns present the number of instances, the number of jobs, and the number of periods. The fourth part indicates the makespan of problems that are calculated by MIP1 and CP1, the fifth part illustrates the elapsed time of two models. The gap for MIP1 is calculated with Equation 3.14. To calculate the improvement of schedules by using CP1 instead of MIP1, we use the average percentage deviation of the best objective line (*ADOL*) of MIP1 against CP1, which is presented in Equation 3.23 and detailed in last column of Table 3.4. (Meyr and Mann 2013).

$$ADOL\% = \frac{MIP_{C_{max}} - CP_{C_{max}}}{MIP_{C_{max}}} \times 100 \qquad (3.23)$$

Table 3.4 shows that MIP1 fails to find the optimal schedules of larger problems. On the other hand, CP1 performs better than MIP1, and it finds successful schedules with an average of 157.6 seconds. In addition, CP1 improves the quality of feasible solutions by 7%, and it is able to generate feasible schedules for larger problems. However, because of the termination limit, CP1 has a heuristic nature, and we evaluate the best limit considering the trade-off between the quality of solutions and the elapsed time in Section 3.4.1.

3.4.1 *Effective Value of the Fail Limit for CP1*

The solution procedure of CP1 is different from MIP1. First of all, CP1 eliminates all infeasible solution points by its propagation techniques and builds the solution space with feasible solution points. Then, it searches the solution space for the best objective value by evaluating each feasible solution point. Therefore, if we limit the CP1 model with a sufficient time, it is able to reach the optimal solution of a problem, and because of this property, the CP1 model can be classified as an exact method. However, the termination limit of models is determined by the user, and this situation adversely affects the quality of the solution. If the user chooses a short time limitation, the model generates a feasible solution quickly, but it may be far from the optimal one. If the user chooses a higher time limitation, the quality of the solution could improve, but the solution time would take longer time. Since the termination limit is a significant parameter for the CP1 model, we need to determine an appropriate time limit by testing the algorithm on different problems.

We generate 10 different problems by using the real production data for past 10 weeks. We consider 1-week schedule as one problem, and we produce 10 different problems, which are described in Section 3.6. We use the fail limit and the elapsed time as a termination limit for CP1. We run the CP1 model with different fail limits and with 1800 seconds time limit. We calculate the *ADOL%* of C_{max} values CP1 models with fail limits against CP1 models with 1800 seconds. Thus, we try to find the gap between the solutions with fail limits and 1800 seconds. We detail our findings in Table 3.5 and in Table 3.6. The first two columns in Table 3.5 give information about the problems, and the remaining columns illustrate the *ADOL%* of C_{max} values between the CP1 models with different fail limits and the CP1 model with time limit. For example, if we use the fail limit

Table 3.5 *ADOL%* Values of CP1 Models with Different Fail Limits against CP1 Models with 1800 seconds

		ADOL% FOR DIFFERENT FAIL LIMITS							
PROBLEM	NO. OF JOBS	CP1_ 100,000	CP1_ 300,000	CP1_ 500,000	CP1_ 700,000	CP1_ 900,000	CP1_ 1,000,000	CP1_ 1,500,000	CP1_ 1800 S
1	63	−12.4	−8.1	−5.8	−3.0	−2.3	−2.3	−1.7	0
2	64	−14.6	−8.5	−6.6	−6.6	−6.6	−6.6	−4.2	0
3	55	−11.8	−5.7	−5.7	−5.7	−5.7	−5.7	−5.7	0
4	61	−13.3	−7.5	−5.8	−5.1	−5.1	−5.1	−2.7	0
5	59	−6.8	−2.4	−2.3	−1.2	−1.2	−1.2	0.0	0
6	58	−10.4	−5.9	−5.9	−0.6	0.0	0.0	0.0	0
7	57	−27.7	−20.2	−7.2	−7.2	−4.0	−2.3	−1.5	0
8	54	−9.8	0.0	0.0	0.0	0.0	0.0	0.0	0
9	54	−15.2	−8.8	−6.6	−2.8	−2.8	−2.8	0.0	0
10	49	−10.0	−6.1	−4.6	−1.3	−1.1	0.0	0.0	0
	Avg.	−13.2	−7.3	−5.0	−3.4	−2.9	−2.6	−1.6	0

Table 3.6 Elapsed Time in Seconds of CP1 Models with Different Termination Limits

		ELAPSED TIME (S)							
PROBLEM	NO. OF JOBS	CP1_ 100,000	CP1_ 300,000	CP1_ 500,000	CP1_ 700,000	CP1_ 900,000	CP1_ 1,000,000	CP1_ 1,500,000	CP1_ 1800 S
1	63	31.5	113.2	199	320.4	578.4	640.5	981.9	1800
2	64	35.7	131.0	262.3	522.4	659.0	702.2	1055.8	1800
3	55	57.8	172.1	258.8	435.6	569.0	609.7	943.3	1800
4	61	40.3	108.4	164.9	222.4	287.0	319.9	450.6	1800
5	59	65.6	187.4	306.6	546.6	578.9	621.0	898.2	1800
6	58	56.5	235.7	379.1	549.6	674.6	756.1	1206.6	1800
7	57	24.9	78.5	108.8	157.8	118.3	247.5	427.8	1800
8	54	54.4	155.6	259.5	393.9	466.4	523.1	817.7	1800
9	54	20.2	47.7	76.8	140.1	227.2	268.9	424.1	1800
10	49	28.4	87.3	195.1	241.5	305.3	393.2	696.1	1800
	Avg.	41.5	131.7	221.1	353.0	446.4	508.2	790.2	1800

as 100,000, we generate schedules with an average of 13.2% worse than the schedules generated in 1800 seconds. Similarly, if we use 700,000 fail limit, the CP1 model generates 3.4% worse schedules than the CP1 model with 1800 seconds, and it only spends 353 seconds on average to reach these solutions instead of 1800 seconds. However, if we choose 900,000 fail limit, the average solution time would increase to 446.4 seconds, but the quality of the solution would only increase by 0.5% on average. Considering the trade-off between the quality of solutions and the elapsed time, we chose the termination limit as 700,000 fail limit.

The results so far show that MIP1 is able to find optimal schedules for small-size instances, but it fails to find feasible solutions in acceptable times for larger instances. On the other hand, CP1 is able to find successful schedules for large instances however, it fails to prove optimality. In conclusion, we see that these models have weaknesses and strengths. We next develop methods to improve both MIP1 and CP1.

3.5 Model Improvements

3.5.1 MIP1 Improvements

Although MIP1 works slow for real-world problems, it provides information on the quality of a solution by either proving optimality or providing an optimality gap. Therefore, we attempt to accelerate the MIP1 model by searching the effects of initiating the model with a lower bound. We convert MIP1 to a relaxed-LP model (LP) and use its objective function value as a lower bound for MIP1. MIP1 starts with this lower bound and tries to find the optimal solution with the same constraints as given in Section 3.3 (MIP1). The experimental results, provided in Table 3.7, indicate that the LP solutions are too far from the optimal solutions, and they would not give a useful lower bound for MIP1.

As a second approach, we use CP1 solution as an initial point for MIP1. CP1 is able to find successful solutions in short times, but it could not prove optimality. We first run CP1 to find a solution, and then we give this solution to MIP1 as a starting point. The idea is that starting a good initial point would increase the performance of

Table 3.7 Comparison of Relaxed LP, MIP1, and CP1 Using Real Data

PROBLEM	NO. OF JOBS	NO. OF PERIODS	C_{MAX} (H)			ELAPSED TIME (S)		
			LP	MIP1	CP1	LP	MIP1	CP1
1	5	5	11.7	16.3[a]	16.3[a]	0.1	1.1	2.9
2	10	8	19.0	29.3[a]	29.3[a]	0.1	6.7	2.7
3	15	13	18.9	44.9	**43.9**	1.0	1800	109.7
4	20	22	18.9	77.6	**73.8**	1.7	1800	119.5
5	25	25	18.9	85.0	**78.8**	2.0	1800	136.1
6	30	27	45.5	99.0	**89.0**	2.5	1800	141.8
7	35	32	45.5	119.2	**98.2**	3.1	1800	175.2
8	40	40	45.5	146.8	**126.1**	8.2	1800	211.0
9	45	49	48.4	—	**156.4**	40.7	1800	305.7
10	52	54	48.3	—	**170.0**	75.8	1800	371.2

[a] Proven optimality.
Note: Highlighted lower C_{max} values are in bold.

MIP1; hence, we may be able to find optimal solutions of large-size problems. We also compare our findings with pure MIP1 solutions, and we record the solution stages of ILOG CPLEX 12.4. Results show that MIP1 with initial solution starts with a smaller gap, but the CPLEX algorithm generates more cuts for pure MIP1 automatically. In conclusion, the initial solution of CP1 is not useful for MIP1.

In this section, we develop different approaches to accelerate MIP1 but none of them is good enough to increase the performance. Therefore we next continue with CP1 and work on improving it.

3.5.2 CP1 Improvements

As mentioned before, CP1 is able to find successful solutions in short times, but it could not prove optimality. The idea is to improve the performance of CP1 so that it may be faster to reach an optimal solution. The CP allows the user to define specific *search algorithms*, which includes rules for variable and value selection in constraint propagation phase. We develop various search algorithms and evaluate their performances. We consider x_{im} decision variable representing the alternative machines of jobs. Intuitively, the jobs with the least alternative machine should be considered first. With this variable selection, we generate three different value selection methods: assigning jobs to machines with the smallest machine index first, assigning jobs to machines with the largest machine index first, and assigning jobs to machines with random order. We compare the performance of these search algorithms considering the solution points, number of branches, and choice points. The experimental results for randomly chosen problems indicate that the number of branches and choice points are the same for three distinct search algorithms. In addition, we increase the number of branches with an average of 85.6% and the choice points with an average of 88.4%, and we decrease the quality of solutions with an average of 0.2% by using proposed search algorithms instead of CP1 with automatic search algorithm. We conclude that the pure CP1 model searches the solution space faster than our search rules, and it reaches better results within same limits. In addition, we consider the most affected jobs in terms of the number of operations and the amount of production. We generate an algorithm that assigns the most affected job first but there is no significant difference between the previous algorithms.

Although we could not improve the performance of CP1 by different search strategies, we could reduce the search space including some constraints to the model. If we aim to minimize the waste time of machines, CP1 would be able to find better schedules. We define a new decision variable called $occupied_{mk}$, which is dependent on machine and period. This variable checks machine m at time $k \times 3600 \times 4$. If machine m is processing a job at that time, the $occupied_{mk}$ is equal to 1, otherwise is 0. This inspection is done for each machine and each period. We also can break the symmetry using constraint (3.24), which indicates that the occupancy rate of the first half of production horizon should be equal to or greater than the second half of production horizon:

$$\sum_{m=1}^{M}\sum_{k=1}^{K/2} occupied_{mk} \geq \sum_{m=1}^{M}\sum_{k=1+K/2}^{K} occupied_{mk} \tag{3.24}$$

To observe the impact of constraint (3.24), we repeat the experiments with randomly generated data that are detailed in Table 3.8. The first part represents the results of CP1 model, and the second part represents the results of CP1 model with constraint (3.24) and the $occupied_{mk}$ decision variables. The new decision variable expands the number of branches and the choice points as expected, but this addition adversely

Table 3.8 Impact of Constraint (3.24) to CP1 Model

	CP1					CP1_CONSTRAINT (3.24)					ADOL%		
NO. OF JOBS	SOLUTION TIME	C_{MAX}	SOLUTION POINTS	NO. OF BRANCHES	CHOICE POINTS	SOLUTION TIME	C_{MAX}	SOLUTION POINTS	NO. OF BRANCHES	CHOICE POINTS	NO. OF BRANCHES	CHOICE POINTS	C_{MAX}
10	2.8	131.3[a]	1	25,259	14,703	4.4	131.3[a]	1	25,433	16,350	−0.7	−11.2	—
15	36.1	259.9	3	238,073	141,762	92.1	259.9	4	313,010	215,426	−31.5	−52.0	—
20	73.5	415.8	1	254,126	156,923	376.4	415.8	4	394,509	297,572	−55.2	−89.6	—
25	99.1	468.8	6	242,452	143,672	456.3	468.8	7	365,114	267,207	−50.6	−86.0	—
30	143.8	**551.0**	9	264,249	166,401	366.4	551.3	8	293,161	194,710	−10.9	−17.0	−0.1
35	191.9	636.9	7	257,140	158,066	667.6	637.9	8	305,326	206,781	−18.7	−30.8	−0.1
40	267.2	**770.2**	4	285,566	187,026	865.6	771.9	3	311,744	212,921	−9.2	−13.8	−0.2
45	313.6	909.8	6	253,935	154,073	1114.5	**906.5**	11	323,773	225,204	−27.5	−46.2	0.4
50	494.6	**982.5**	7	280,986	180,449	1833.8	984.8	20	321,911	222,315	−14.6	−23.2	−0.2
										Avg.	−24.3	−41.1	−0.1

[a] Proven optimality.

Note: Highlighted lower C_{max} values are in bold.

affects the performance of CP1 with constraint (3.24). Pure CP1 finds better schedules within the same limits. According to the generated schedules by two models, it is observed that there is no difference between solutions in terms of occupancy rates. In conclusion, we can claim that CP1 generates schedules with the same logic without constraint (3.24).

The symmetry breaking is an important tool to reduce the solution space of CP1. Using an effective symmetry-breaking constraint decreases the alternative solutions and improves the performance of the model. In the previous experiment, we attempt to reduce the solution space by constraint (3.24) considering the time axis, which is already performed by pure CP1. However, we observe that in the optimal schedules, the first half of the machines are more occupied than the other machines. Therefore, we generate a new symmetry-breaking constraint (3.24′) instead of constraint (3.24), which takes into account the symmetry of assigned jobs to machines:

$$\sum_{a \text{ in } Alter|a.m \leq 6} presenceOf(x_a) \geq \sum_{a \text{ in } Alter|a.m > 6} presenceOf(x_a) \quad (3.24')$$

To investigate the impact of constraint (3.24′), we repeat the experiments and present the findings in Table 3.9.

According to the last part of Table 3.9, if we use CP1_constraint (3.24′) instead of CP1, we can reduce the number of branches with an average of 1.7%, choice points with an average of 1.6%, and improve the quality of solutions with an average of 0.04% by reaching more feasible solutions. In conclusion, we can improve the performance of CP1 including the symmetry-breaking constraint (3.24′) to the model.

In this section, we develop methods to improve the performance of MIP1 and CP1. Although we could not achieve an improvement with MIP1, we accelerate CP1 by constraint (3.24′). We next compare the performance of CP1 with the real schedules.

3.6 Comparison of Real-World Schedules with CP1

Beycelik Gestamp works with weekly demands. A production engineer spends the first day of the week to generate schedules and spends 4 hours a day to make sure the production is running smoothly.

Table 3.9 Impact of Constraint (3.24′) to CP1 Model

	CP1					CP1_CONSTRAINT (3.24′)					ADOL%		
NO. OF JOBS	SOLUTION TIME	C_{MAX}	SOLUTION POINTS	NO. OF BRANCHES	CHOICE POINTS	SOLUTION TIME	C_{MAX}	SOLUTION POINTS	NO. OF BRANCHES	CHOICE POINTS	NO. OF BRANCHES	CHOICE POINTS	C_{MAX}
10	2.8	131.3	1	25,259	14,703	3.0	131.3	1	22,231	12,942	12.0	12.0	0
15	36.1	259.9	3	238,073	141,762	39.9	259.9	2	249,220	154,250	−4.7	−8.8	0
20	73.5	415.8	1	254,126	156,923	74.5	415.8	1	254,159	156,949	0.0	0.0	0
25	99.1	468.8	6	242,452	143,672	96.8	468.8	6	240,544	142,679	0.8	0.7	0
30	143.8	551.0	9	264,249	166,401	131.9	551.0	10	257,615	160,432	2.5	3.6	0
35	191.9	636.9	7	257,140	158,066	178.4	636.9	7	254,668	155,192	1.0	1.8	0
40	267.2	770.2	4	285,566	187,026	240.4	770.2	4	285,798	187,381	−0.1	−0.2	0
45	313.6	909.8	6	253,935	154,073	319.2	**906.5**	5	266,882	167,312	−5.1	−8.6	0.36
50	494.6	982.5	7	280,986	180,449	451.4	**981.8**	15	255,443	155,303	9.1	13.9	0.07
										Avg.	1.7	1.6	0.05

Note: Highlighted lower C_{max} values are in bold.

Thus, the engineer spends 28 hours to maintain the production for the whole week. However, we can generate successful schedules in minutes by CP1. In order to simulate the real-world schedules, we examine the real-world schedules of past 10 weeks, and we generate 10 different problems. We consider the actual production time in the company by eliminating shift/lunch breaks. The company and CP1 assume that jobs may not be preempted, the production period is 4 hours, and the changeovers are done in shift/lunch breaks. Thus, we do not need to consider the changeovers in both systems. However, the company faces with two groups of interruptions; the unpredictable interruptions include breakdowns of machines and maintenance activities, and the predictable interruptions include educations of operators and public holidays. To reflect the real system accurately, we consider these interruptions as dummy jobs. We first determine the start time, the duration, and the related machine to which interruption happened. We run the CP1 model without interruptions. If the model generates a schedule that is completed before the interruption, we do not need to consider the interruption. For example, if a model finds a schedule with the $C_{\max}$ value of 100 hours and a breakdown happens after 120 hours, we do not need to consider this interruption. On the other hand, if the length of the schedule includes the start time of the interruption, we define a dummy job to represent the interruption. The dummy job includes only one operation and only one alternative machine. The processing time is determined by considering the total duration of the interruption. We run the CP1 model with dummy jobs and with a new constraint to present the interruptions at the same time and machine with the real-world schedule. If an interruption occurs at machine m in period k, the dummy job i' is represented with the constraint (3.25) in CP1 model:

$$startOf(x_a) = k \times 4 \times 3600, \quad \forall a \in Alter|\ a.m = m,\ a.i = i' \qquad (3.25)$$

The dummy jobs have the same durations with the real-world schedules, and they are assigned to the same machine at the same time. Therefore, we can reflect the production horizon with interruptions precisely. We run the CP1 model with 700,000 fail limit (CP1_700,000) and 1800 seconds (CP1_1800) as seen in Table 3.10. The first two columns give information about the problems, and

Table 3.10 Comparison of CP1 with Real-World Schedules

PROBLEM	NO. OF JOBS	C_{MAX} (H)			ELAPSED TIME (S)			ADOL%	
		COMPANY	CP_700,000	CP_1800	COMPANY	CP_700,000	CP_1800	CP_700,000	CP_1800
1	63	150	108.9	108.1	100,800	416.4	1800	27.4	28.0
2	64	142.5	120.9	116.5	100,800	280.0	1800	15.2	18.2
3	55	142.5	110.4	105.5	100,800	338.0	1800	22.5	25.9
4	61	150	121.6	119.3	100,800	374.8	1800	18.9	20.3
5	59	142.5	112.3	111.[illegible]	100,800	394.4	1800	21.2	22.1
6	58	142.5	116.7	112.[illegible]	100,800	570.8	1800	18.1	21.3
7	57	150	118.1	118.1	100,800	225.6	1800	21.3	21.3
8	54	127.5	106.7	106.7	100,800	232.1	1800	16.3	16.3
9	54	135	134.2	131.8	100,800	358.4	1800	0.6	2.3
10	49	150	137.0	137.0	100,800	374.7	1800	8.7	8.7
				Avg.	100,800	356.5	1800	17.0	18.4

the remaining columns illustrate the C_{max} values of schedules by CP1_700,000 and CP1_1800, the elapsed times, and the *ADOL%* values of CP1 models against the real-world schedules of the company. The company spends 28 hours on average to generate the real-world schedules. However, we can improve the manually generated schedules with an average of 17% by CP1_700,000, and the model spends only 356.5 seconds instead of 28 hours (100,800 seconds). If we aim to find better schedules, we can improve the schedules with an average of 18.4% by using CP1_1800. Although CP1_1800 is able reach better results, the difference between CP1_700,000 and CP1_1800 is 1.4% on average, and the company may choose to generate weekly schedules in 356.5 seconds rather than 1800 seconds. In conclusion, the CP1 model is able to improve the weekly schedules with an average of 17%, which has a positive effect on reducing the production cost and increasing the weekly capacity.

3.7 Conclusion

This chapter presents a novel problem description of stamping scheduling. The problem is presented by a MIP model utilizing the conventions and practices of the company. The complexity of the problem adversely affects the quality of solutions, and MIP1 could not find any feasible solutions of the real-world instances of the problem. We next develop a new model, CP1, by using CP. The CP1 model is able to find successful solutions of the problem instances; however, its termination limit has a negative effect on the solutions. To determine the trade-off between the solution quality and the solution time, we run experiments to determine a good termination limit. Although we develop various approaches to combine MIP1 and CP1, the pure CP1 model outperforms MIP1 and combined methods. Therefore, we focus on improving CP1 model by considering the symmetry-breaking constraints. We also compare CP1 model with manually generated schedules by the company. The CP1 model improves the weekly schedules with an average of 17% and reaches these solutions within 356.5 seconds. The company can convert CP1 model to a practical software and can use it to generate effective schedules in minutes.

References

Baptiste, P., Le Pape, C., and Nuijten, W. (2001). *Constraint-Based Scheduling: Applying Constraint Programming to Scheduling Problems* (Vol. 39). Springer, Berlin, Germany.

Barlatt, A. Y., Cohn, A., Gusikhin, O., Fradkin, Y., Davidson, R., and Batey, J. (2012). Ford Motor Company implements integrated planning and scheduling in a complex automotive manufacturing environment. *Interfaces*, 42(5), 478–491.

Błażewicz, J., Ecker, K. H., and Pesch, E. (Eds.). (2007). *Handbook on Scheduling [Electronic Resource]: From Theory to Applications.* Springer, Berlin, Germany.

Edis, E. B., Oguz, C., and Ozkarahan, I. (2013). Parallel machine scheduling with additional resources: Notation, classification, models and solution methods. *European Journal of Operational Research*, 230(3), 449–463.

Edis, E. B. and Ozkarahan, I. (2012). Solution approaches for a real-life resource-constrained parallel machine scheduling problem. *The International Journal of Advanced Manufacturing Technology*, 58(9–12), 1141–1153.

Khayat, G. E., Langevin, A., and Riopel, D. (2006). Integrated production and material handling scheduling using mathematical programming and constraint programming. *European Journal of Operational Research*, 175(3), 1818–1832.

Malapert, A., Guéret, C., and Rousseau, L. M. (2012). A constraint programming approach for a batch processing problem with non-identical job sizes. *European Journal of Operational Research*, 221(3), 533–545.

Meyr, H. and Mann, M. (2013). A decomposition approach for the general lotsizing and scheduling problem for parallel production lines. *European Journal of Operational Research*, 229(3), 718–731.

Nuijten, W. P. and Aarts, E. H. (1996). A computational study of constraint satisfaction for multiple capacitated job shop scheduling. *European Journal of Operational Research*, 90(2), 269–284.

Pinedo, M. (2012). *Scheduling: Theory, Algorithms, and Systems.* Springer, Berlin, Germany.

Relvas, S., Boschetto Magatão, S. N., Barbosa-Póvoa, A. P. F., and Neves Jr., F. (2013). Integrated scheduling and inventory management of an oil products distribution system. *Omega*, 41(6), 955–968.

Van Hentenryck, P. (1999). *The OPL Optimization Programming Language.* MIT Press, Cambridge, MA.

4
Voting Security in Social Media

SEIFEDINE KADRY

Contents

4.1 Introduction

Social networks are websites on the Internet with a common ownership. The members of social networks have some privacy, with some information being available for selected people but blocked for others. The security problem is a very complex issue in social networks. Theft of personal data and intrusion into profile pages are recorded in large numbers. Researchers and developers are therefore looking for ways to protect these data. The biggest problem is how to protect the voting in social networks from the risk of penetration. The most important risks in voting are votes that are stolen by candidates or those obtained by creating fake votes. In this chapter, we develop a fast and accurate process for checking the social network voting data. This method will compare the votes cast by voters with those obtained by candidates with the use of algorithms to detect any stolen or fake votes [1].

4.2 What Are Social Networks?

Social networks are networks of relationships between institutions or individuals that share a common bond, such as classmates, coworkers, or family members. Social networking is the number of websites providing services to users for blogs, e-mail, publishing articles, posting photos, exchanging ideas, advertising, etc. [2].

The structure of social networks makes it easier for researchers to visualize the structure of social relations between institutions and individuals.

In the last few years, social networking sites have increasingly spread around the world at a fast pace. Some social networking websites involve as many as 800 million users. The most popular social networking websites that have fastest and widespread usage worldwide are Wikipedia, Facebook, Twitter, and Myspace, among others.

Wikipedia, deriving from *wiki*, meaning "quick" in Hawaiian, and "pedia," the suffix in "encyclopedia," was launched in January 2001. Wikipedia is a free encyclopedia where any user can create or edit articles. It is a very important website on the Internet comprising a classification of terms and is the sixth most visited website worldwide. In 2012, Wikipedia was available in more than 225 languages. In the United States alone, the Wikipedia website recorded 2700 million visits per month [3].

Social network analysis (SNA) is a method of viewing the structure of social networks, where each individual or institution within the social network is considered a node and every relationship between the nodes is called a link (edge). The purpose of graphical social networks is to understand the structure and the links between members, and to apply social theories to them [2].

Voting on social networks is one of the main reasons for the success of the idea of social networks on the Internet. Social networks allow users the freedom to express their opinions and vote democratically. Users of social networks can vote for an article, image, or video [4].

For example, they can vote for a video on YouTube (like or dislike) or a status or photo on Facebook (like); shortly, Facebook will be adding "dislike" to voting. For Wikipedia, one can vote to choose administrators (negative or positive). As Wikipedia is a nonprofit website, administrators are selected by the election of some incompetent users. The duty of administrators is correcting articles.

The most important goal of voting on social networks is to know the views of the members of the community and their aspirations and promotion of social democracy.

4.3 Network Workbench Tool

The network workbench (NWB) tool is the most famous and the latest SNA software in recent years. The NWB (Figure 4.1) is a network analysis, modeling, and visualization toolkit for physics, biomedical, and social science research [5–7].

The NWB is designed on a Cyber Infrastructure Shell (CIShell). In 2007, it became an open-source software framework for the easy integration and utilization of datasets, algorithms, tools, and computing resources. The NWB defines the features of a social network in the simplest way, through representation by numbers, graphs, and histograms.

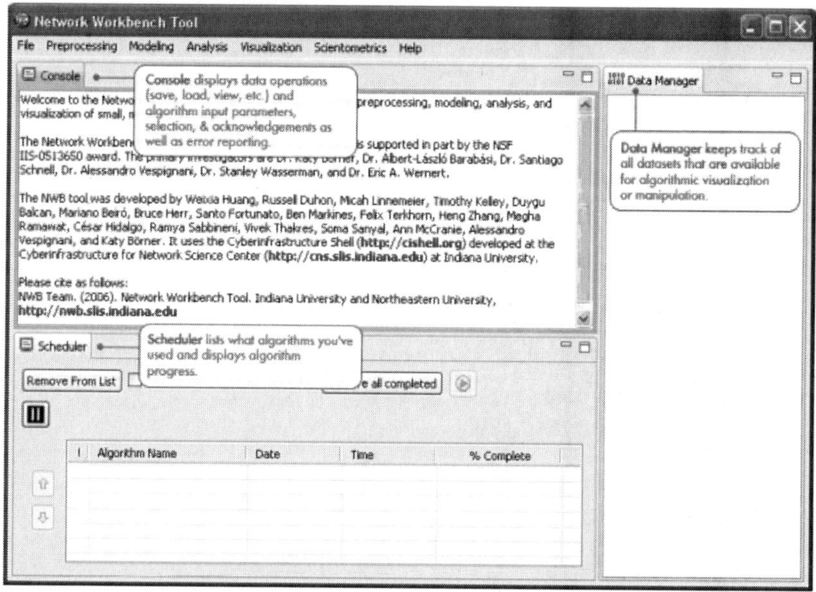

Figure 4.1 User interface for NWB.

The NWB is an easy tool for researchers and developers of the concept of social networks that involve a large number of algorithms to be applied to extensive data of social networks. This tool allows users to create a model network and apply the necessary algorithms; they may also use different visualizations to represent the network to effectively adopt the characteristics of the network analyzed.

In 2008, the NWB was updated as having 80 algorithms and 30 sample datasets. As the NWB tool is developed using JAVA, algorithms developed in other programming languages such as FORTRAN, C, and C++ can be easily integrated. The NWB interface is easy to use and consists of several lists containing the names of the algorithms, of network models, and a number of other functions available in the tool.

4.4 Simulation and Results

4.4.1 Data Description

Wikipedia is a nonprofit organization that contains a large number of articles, which continue to increase (in 2012, over 3.9 million articles were published in English). Articles are constantly added by the numerous users of this website. A small number of users are administrators, who are responsible for the maintenance and amendment of articles. The administrators of Wikipedia must be elected by its users. The data shown in Figure 4.2 are taken from January 2008. These data users vote to select administrators [8–10].

Users have the right to vote for positive or negative. As shown in Figure 4.3, the total number of votes is 103,663 and the number of users participating in the voting is 7,066. The group comprises users and administrators. The number of people elected is 2794 (i.e., those who received positive and negative votes). Of these, there were 1235 winners who received votes in their favor and were promoted as administrators. The 1559 losers received negative votes.

4.4.2 Voting Security

The in-degree of a node is an edge (or link) coming to the node. The algorithm builds two histograms, each showing the values of in-degree of all nodes calculated in a different way. In the first histogram

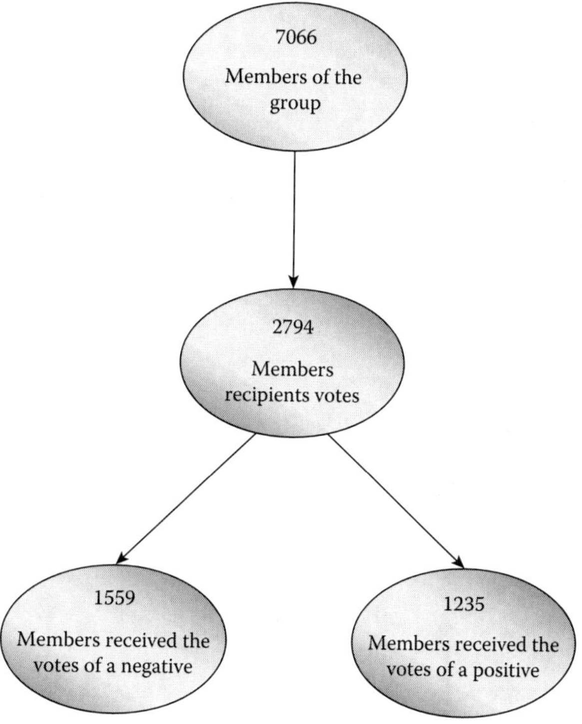

Figure 4.2 Structure of data.

(Figure 4.4), the occurrence of any in-degree value between the minimum and the maximum is estimated and divided by the number of nodes in the network in order to obtain the probability. In the second histogram (Figure 4.5), the interval spanning the values of in-degree is divided into bins, whose size grows while moving toward higher values of the variable. The size of each bin is obtained by multiplying the size of the previous bin by a fixed number [11,12].

The purpose of using this algorithm to the data is to know the probability of the votes received by the candidates, where each node in the network is considered a candidate and each edge received by a node is a vote.

The out-degree of a node is an edge (link) going out of a node. The application of the out-degree algorithm gives two histograms, each showing the values of out-degree of all nodes calculated in a different way. In the first histogram (Figure 4.6), the occurrence of any out-degree value between the minimum and the maximum is

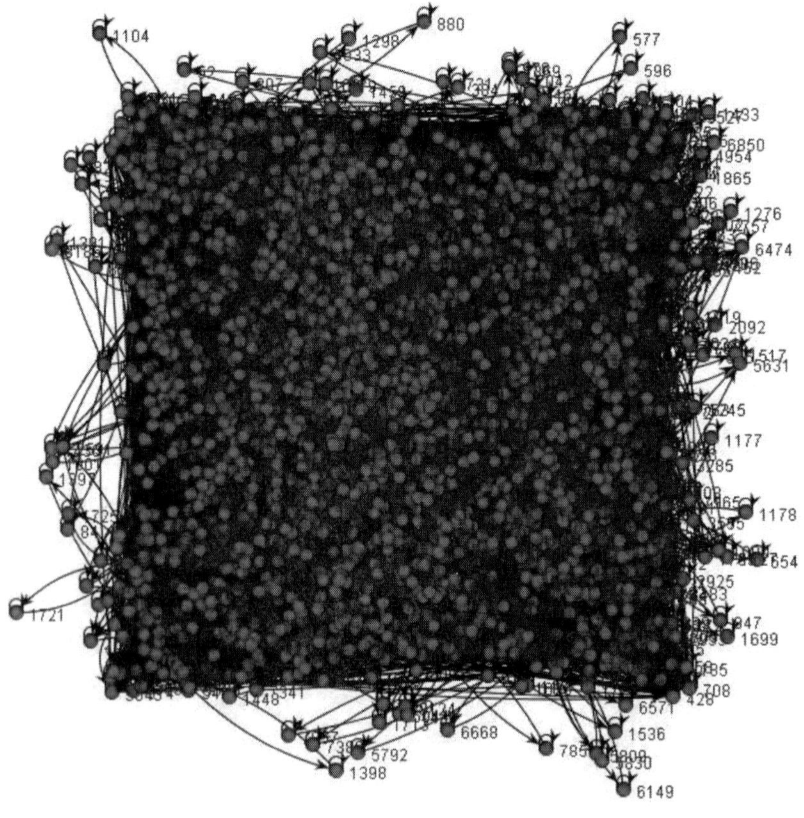

Figure 4.3 Extract of data network.

estimated and divided by the number of nodes in the network to obtain the probability. In the second histogram (Figure 4.7), the interval spanning the values of out-degree is divided into bins whose size grows while moving toward higher values of the variable. The size of each bin is obtained by multiplying the size of the previous bin by a fixed number [13,14].

The purpose of applying this algorithm to the data is to know the probability of the votes sent from the voters, where each node in the network is considered a voter and each edge sent from the node is a vote.

The PageRank algorithm calculates the rank of web pages. The histogram in Figure 4.7 is divided into intervals spanned by the PageRank values. One divides the range of variation of the PageRank into equal bins and determines the number of nodes whose PageRank

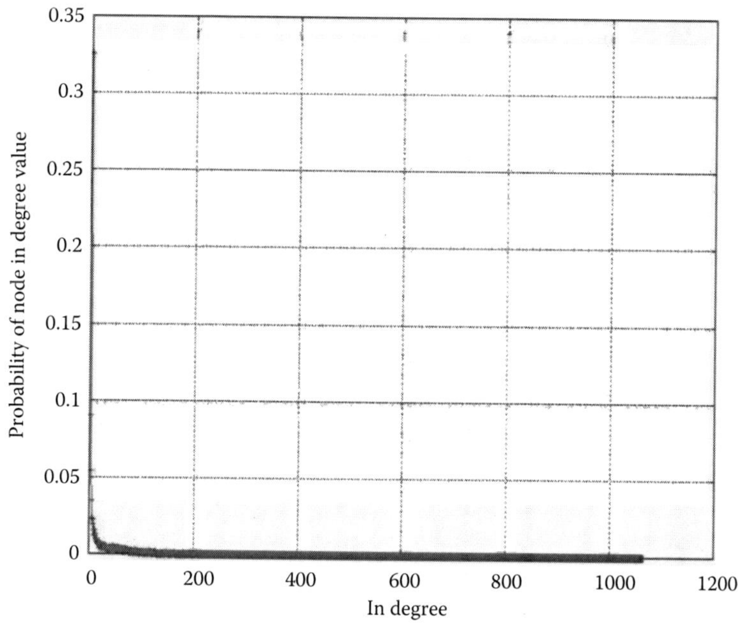

Figure 4.4 In-degree distribution divided by the number of nodes.

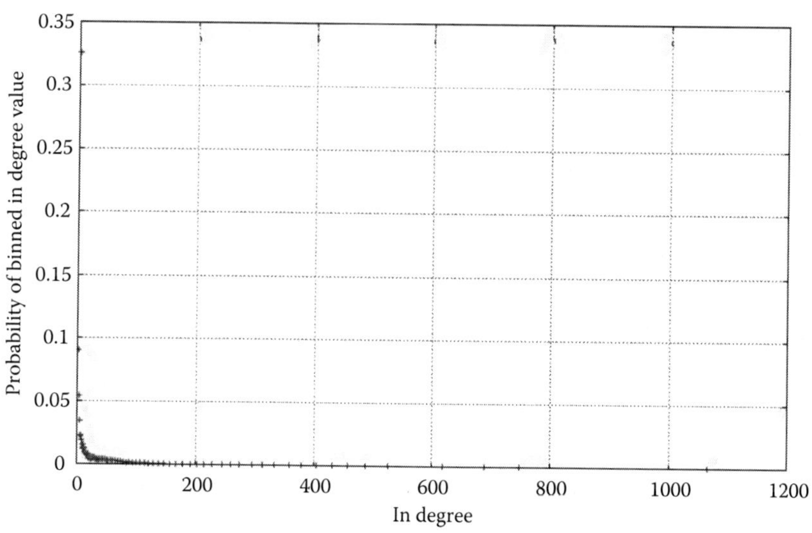

Figure 4.5 In-degree distribution divided by the number of bins.

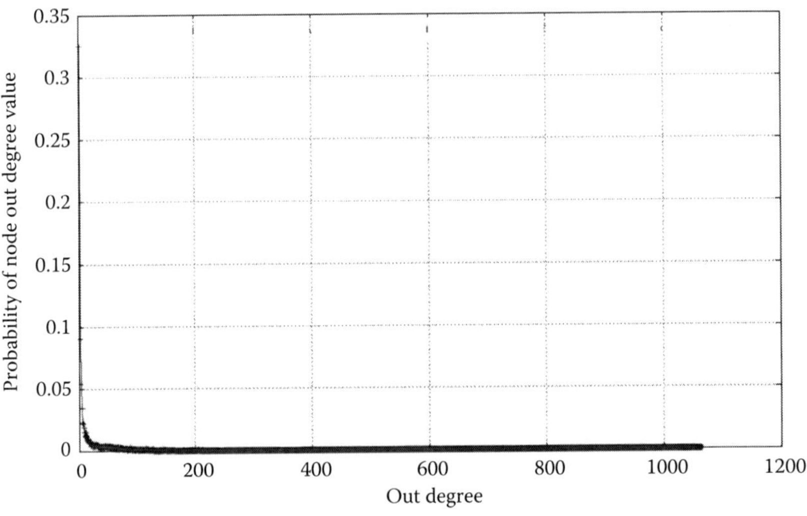

Figure 4.6 Out-degree distribution divided by the number of nodes.

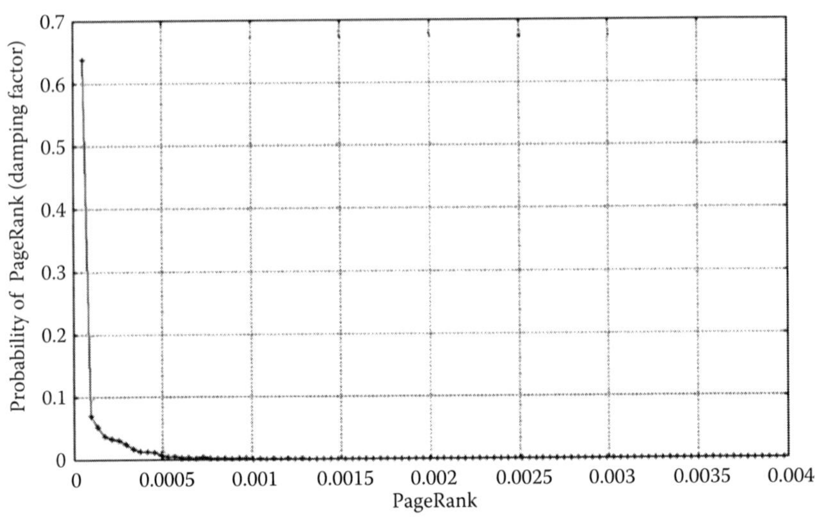

Figure 4.7 In-degree distribution divided by the number of bins.

values lie inside each bin: the scores are then divided by the number of nodes in the network, to obtain the probability. The damping factor is a real number between 0 and 1, which is considered for calculating the probability. When measuring the PageRank of a large website such as Google, the damping factor is set at 0.85. When measuring the PageRank of the data of social networks, the damping factor is usually set at 0.15 (i.e., when we apply the algorithm, we set the value of the damping factor at 0.15). The purpose of the application of this algorithm on the data is to know the probability of the rate of visiting a web page in order to vote [15].

4.5 Heider's Balance Theory

Heider's balance theory is the most important theories of social analysis. This theory explains the relationship between individuals in a social network on the basis of emotions. This theory can be applied to several social platforms, businesses, elections, and even to know those following people on a particular subject [16,17]. In social networks, it can be used to know the structure of the social network, as well as to measure the strength of the relationship and discover the type of relationship between individuals within the social network.

If a balanced state occurs between two people (e.g., if there is a link between the first person A and the second person B such that A → B and B → A), this relationship is called dyad reciprocity. But if the relationship between three people {A, B, C} is such that A → B, B → C, C → A, this is called a triad. Table 4.1 shows the results of applying the algorithms to the data of votes in Wikipedia. Status 1, in which the dyad algorithm is used, calculates the number of dyads with the reciprocated relation A → B and B → A. This number is very large compared with the links between the nodes in the network. In other words, the voting relationship among the users of Wikipedia is too large. This means that there is a good relationship between individuals in the network.

Status 2, in which the triad algorithm is used, calculates the number of ordered triads (A → B and B → C). Heider's balance theory assumes the principle of triangular relationships. For example, my

Table 4.1 Votes in Wikipedia Results

STATUS	ALGORITHM	RESULT
1	Number of dyads with reciprocated relation	100,762
2	Number of ordered triads (A → B and B → C)	29,091,160
3	Number of transitive ordered triads (A → B, B → C, and C → A)	3,650,334

friend's friend is my friend, my friend's enemy is my enemy, my enemy's friend is my enemy, my enemy's enemy is my friend. This means that if user A has a positive link with user B and user B has a positive link with user C, in all probabilities user C will have a positive link with user A. Here, one theoretically guesses the votes issued by user C, although the votes are not yet present.

This theory could even benefit political elections, where one can guess who would choose to vote later either because they are under the legal age for voting or because they did not participate for other reasons. For example, party A obtains 100 votes in the first city and party B obtains 100 votes in the second city. This theory solves the problem of equality of number of votes of the two parties. If the first city has more legal (in terms of age) voters than the second city, then party A is the winner because this corresponds to the ideas, ideology, and religion of the voters and their family members and the population in the city.

In status 3, the triad algorithm calculates the number of transitive ordered triads (A → B, B → C, and C → A). In other words, it takes into account the three links between the nodes that already exist without needing estimation.

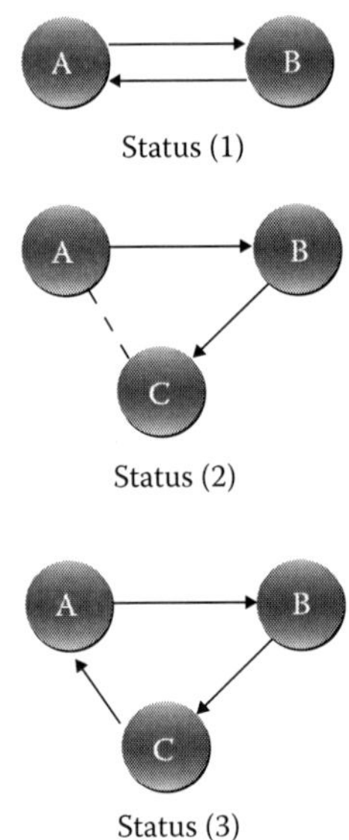

The balance theory enables us to check whether the data in a network are real or imaginary. If the data are obtained from a reliable source or if they are already existing data for a social network, the data are considered real. When the dyad and triad algorithms were applied to network data, the results produced large numbers of links, which indicates that the network data were balanced and real. Many of the links within the network shared the dyad relationship more than the triad.

4.6 Conclusion

The purpose of this study is to check large amounts of data in social networks. If the data are vulnerable to hacking and counterfeiting, this can be detected by using certain algorithms contained within the software of the SNA.

In this chapter, we examine the data for Wikipedia's voting to choose its administrators. Results of the analysis revealed that the election was not exposed to fraud and electronic piracy operations. The NWB tool was used for the analysis. The in-degree and out-degree algorithms were applied to calculate the number of votes cast by the voters and the number obtained by the candidates. If the curve for in-degree is similar to the curve for out-degree, this means that the number of votes cast by the voters equals the number of votes obtained by the candidates. When applying the k-nearest neighbors (KNN) algorithm, the tool compares the degree of nodes with the degree for their neighbors; in other words, the number of votes is compared with the votes for neighbors. In four cases that compared the sent and received votes for users with those for their neighbors, the curves obtained were similar. When applying PageRank to measure the rate of users' access to the voting page, the rate of users logging in to the voting web page was compatible with the number of voters and the number of votes in the network. This indicates that no voted are deleted or added.

The balance theory proved that whether dyad or triad relations were found between the nodes inside a network, there were large numbers of links in the social network. This indicates that the network data are real and not fictional. It also shows the strength of social cohesion between individuals involved in Wikipedia. Moreover, the balance theory also allows the use of these algorithms to predict the votes of some voters.

Researchers in this field are working on expanding the use of these algorithms for a quick check of data, especially for electronic voting in social networks and in other areas. Moreover, there is a need to develop efficient algorithms for the detection of fraud in electronic voting. Voting in social networks is a good and important way of getting feedback from the community about social issues. Thus, social networks play a significant role in granting freedom of expression for members of the community.

References

1. G. Hogben (2009) Security issues in the future of social networking. ENISA Position Paper for *W3C Workshop on the Future of Social Networking*, Barcelona, Spain.
2. D. Passmore (2011) Social network analysis: Theory and applications. Retrieved, January 3, 2011, from http://train.ed.psu.edu/WFED-543/SocNet_TheoryApp.pdf.
3. Wikipedia. Retrieved, June 15, 2012, from http://en.wikipedia.org/wiki/Wikipedia.
4. P. Boldi, F. Bonchi, C. Castillo, and S. Vigna (2009) Voting in social networks. In *Conference on Information and Knowledge Management (CIKM)* (2009), pp. 777–786.
5. Network Workbench About. Retrieved, June 15, 2012, from http://nwb.cns.iu.edu/about.html.
6. B. Herr and W. Huang (2007). Introduction to the network workbench and cyber infrastructure shell, Introductory Paper. School of Library and Information Science, Indiana University, Bloomington, IN.
7. B. (Weixia) Huang, M. Linnemeier, T. Kelley, and R. J. Duhon (2009) Network workbench tool. Cyber infrastructure for Network Science Center, School of Library and Information Science, Indiana University, Bloomington, IN.
8. Stanford University. Wikipedia vote network. Retrieved, June 15, 2012, from http://snap.stanford.edu/data/wiki-Vote.html.
9. J. Leskovec, D. Huttenlocher, and J. Kleinberg (2010) Predicting positive and negative links in online social networks, ACM WWW International Conference on World Wide Web (WWW).
10. J. Leskovec, D. Huttenlocher, and J. Kleinberg (2010) Signed networks in social media, ACM SIGCHI Conference on Human Factors in Computing Systems (CHI).
11. Network Workbench. Node in-degree. Retrieved, June 15, 2012, from https://nwb.slis.indiana.edu/community/?n = AnalyzeData.NodeIndegree.

12. Network Workbench. In-degree distribution. Retrieved, June 15, 2012, from https://nwb.slis.indiana.edu/community/?n = AnalyzeData.IndegreeDistribution.
13. Network Workbench. Node out-degree. Retrieved, June 15, 2012, from https://nwb.slis.indiana.edu/community/?n = AnalyzeData.NodeOutdegree.
14. Network Workbench. Out-degree distribution. Retrieved, June 15, 2012, from https://nwb.slis.indiana.edu/community/?n = AnalyzeData.OutdegreeDistribution.
15. Network Workbench. PageRank. Retrieved, June 15, 2012, from https://nwb.slis.indiana.edu/community/?n = AnalyzeData.PageRank.
16. D. Khanafiah and H. Situngkir (2003) Social balance theory revisiting Heider's balance theory for many agents. Technical Report. Retrieved, July 11, 2014, from http://cogprints.org/3641/1/Heidcog.pdf.
17. N. P. Hummon and P. Doreian (2003) Some dynamics of social balance processes: Bringing Heider back into balance theory, *Social Network* 25: 17–49.

5

Integrated Approach to Optimize Open-Pit Mine Block Sequencing

AMIN MOUSAVI, ERHAN KOZAN, AND SHI QIANG LIU

Contents

5.1 Introduction

Nowadays, optimization techniques have been applied to solve a variety of problems in mining industries for production planning, production scheduling, determining capacities, optimizing mining layout, obtaining optimal resource allocation, determining material destinations, equipment maintenance, and rostering (Kozan and Liu, 2011; Newman et al., 2010; Osanloo et al., 2008).

One of the vital optimization problems in open-pit mining is to determine the optimal extraction sequences of material. The open-pit mine sequencing problem is defined as specifying the sequence in which material should be extracted from pits and then transferred to appropriate destinations. Generally, material with no economic value is dumped while profitable material is processed at mills or stocked at stockpiles for future usage.

In the first step of mining operation optimization, mineral deposit is divided into several blocks, and attributes such as grade and density are estimated for each block. The set of these blocks for whole deposit and surrounding area is called three-dimensional (3D) block model. A block model contains several thousands to over several millions of blocks based on the size of the orebody and the dimensions of each individual block. The block model provides most important information for open-pit optimization problems. Kriging estimator as the best linear unbiased estimator and geo-statistical simulation methods such as sequential Gaussian simulation, p-field simulation, and simulated annealing are widely used to estimate block attributes (Lark et al., 2006; Vann et al., 2002; Verly, 2005). After estimating block characteristics, block economic value or cash flow of the block can be calculated by considering economical parameters.

Identifying the grade that classifies material into waste or ore has been a challenging subject in mining engineering for several decades. A cut-off grade is applied to distinguish ore and waste in a given mineral deposit. Generally, cut-off grade is defined as the minimum amount of valuable mineral that must exist in one unit (e.g., one tone) of material before this material is sent to the processing plant. In conventional methods, cut-off grade is determined as a function of price of product, cash costs of mining and processing, while in reality, capacities of mining and processing as well as the grade-tonnage distribution of the deposit should be taken into account (Asad and Topal, 2011; Johnson et al., 2011). Therefore, the most comprehensive method to determine dynamic cut-off grade is to integrate cut-off grade optimization with the determination of the extraction sequencing.

The mine production sequencing problem may be solved for different levels of the accuracy. For the simplification reasons, some blocks are aggregated into bigger units to obtain extraction sequences of these units with less computational efforts (Askari-Nasab et al., 2010; Ramazan, 2007). As a result of the aggregation, the distinct nature of these blocks is ignored. Therefore, to keep the resolution of solution, the production sequencing problem is solved at the level of block, smallest unit of material of which attributes are estimated, and the problem is called open-pit block sequencing (OPBS) problem (Cullenbine et al., 2011). The OPBS is a challenging problem to be solved due to the size of problem in terms of the number of decision variables and constraints, and the complexity of the problem. Typically, the constraints relate to accessibility to the blocks, mining and milling capacities, grades of mill feed and concentrates, capacities of extraction equipment, and physical and operational requirements such as minimum required width for the machinery. The objective of the OPBS problem is to maximize mining economic value with mining operation efficiency.

In the literature, mixed integer programming (MIP) was used to formulate the OPBS problem. Dagdelen and Johnson (1986) applied Lagrangian relaxations to relax and solve the MIP model of the OPBS problem. Bienstock and Zuckerberg (2010) presented a column generation method with some modifications in iterations to solve linear programming relaxations. Ramazan (2007) proposed a tree algorithm to reduce the size of MIP formulation. Caccetta and Hill (2003) presented a branch and cut algorithm to solve the model, but the details were not given in their paper. Ramazan and Dimitrakopoulos (2004) proposed a relaxed MIP model with fewer binary variables. Menabde et al. (2004) presented a MIP model that integrates cut-off grade simultaneously. Boland et al. (2007) presented a disaggregation method in order to control the processing feed at the level of block decision and heighten the variable freedom.

To solve the OPBS problem more efficiently, several authors developed heuristics. Gershon (1987) proposed a heuristic approach for the OPBS. In this approach, blocks are ranked based on the value of blocks that are located beneath the given block. Blocks with higher rank have priority to be extracted. Tolwinski and Underwood (1996) combined dynamic programming, stochastic optimization, artificial intelligence, and heuristic approaches to solve the OPBS problem. Cullenbine et al. (2011) proposed sliding-time-window heuristic algorithm to solve the model. Chicoisne et al. (2012) combined LP relaxations of the problem with a topological-sorting-based rounding algorithm. Sattarvand and Niemann-Delius (2008) discussed metaheuristics that were applied to the OPBS. In an early work, Denby and Schofield (1994) applied genetic algorithm to solve the large-size problem. Kumral and Dowd (2005) recommended simulated annealing to obtain the solution of the OPBS problem. The Lagrangian relaxations technique was applied to obtain a suboptimal solution, and later simulated annealing was applied to improve the initial solution of the multiobjective model. Ferland et al. (2007) developed a hybrid greedy randomized adaptive search procedure (GRASP) and particle swarm algorithm. The GRASP was employed to construct the initial population (swarm). Then, particle swarm algorithm searches within a feasible domain to improve the initial solution. Sattarvand (2009) presented an algorithm based on ant colony optimization (ACO) algorithm. Myburgh and Deb (2010) presented an evolutionary algorithm based on predeterministic stripping ratio to solve large-sized OPBS and introduced evORElution package. Lamghari and Dimitrakopoulos (2012) proposed Tabu search algorithm to solve OPBS problem. In order to investigate extensive domain, they applied long-term memory and variable neighborhood search methods.

In this chapter, a model with real-life constraints is developed for the OPBS problem. This model will be applicable to find the optimum extraction sequence of blocks over the hourly based periods. The computational experiments are performed to validate the proposed model and to recommend the solution approaches for the real-size cases.

5.2 Mathematical Programming

A general process flow in an iron ore mine is shown in Figure 5.1. Blocks are extracted by excavators, and the mined material is carried to different destinations by trucks. This run of mine material is classified into the waste, low, or high grade based on the block content. Stockpiles are designed for mixing or blending material with different characteristics or to defer the processing of material to future. The low-grade material is enriched in the processing plants to achieve the final product requirements. Finally, the product of low-grade plants and high-grade crusher is blended and sent to the rail loading area as the final product.

Based on the analysis in Figure 5.1, the OPBS problem aims to determine the optimal sequence of blocks. After solving block sequencing problem, two important questions will be answered: which blocks are selected to be extracted over periods, and which destination will each block be sent to?

The problem is formulated using the following notations.

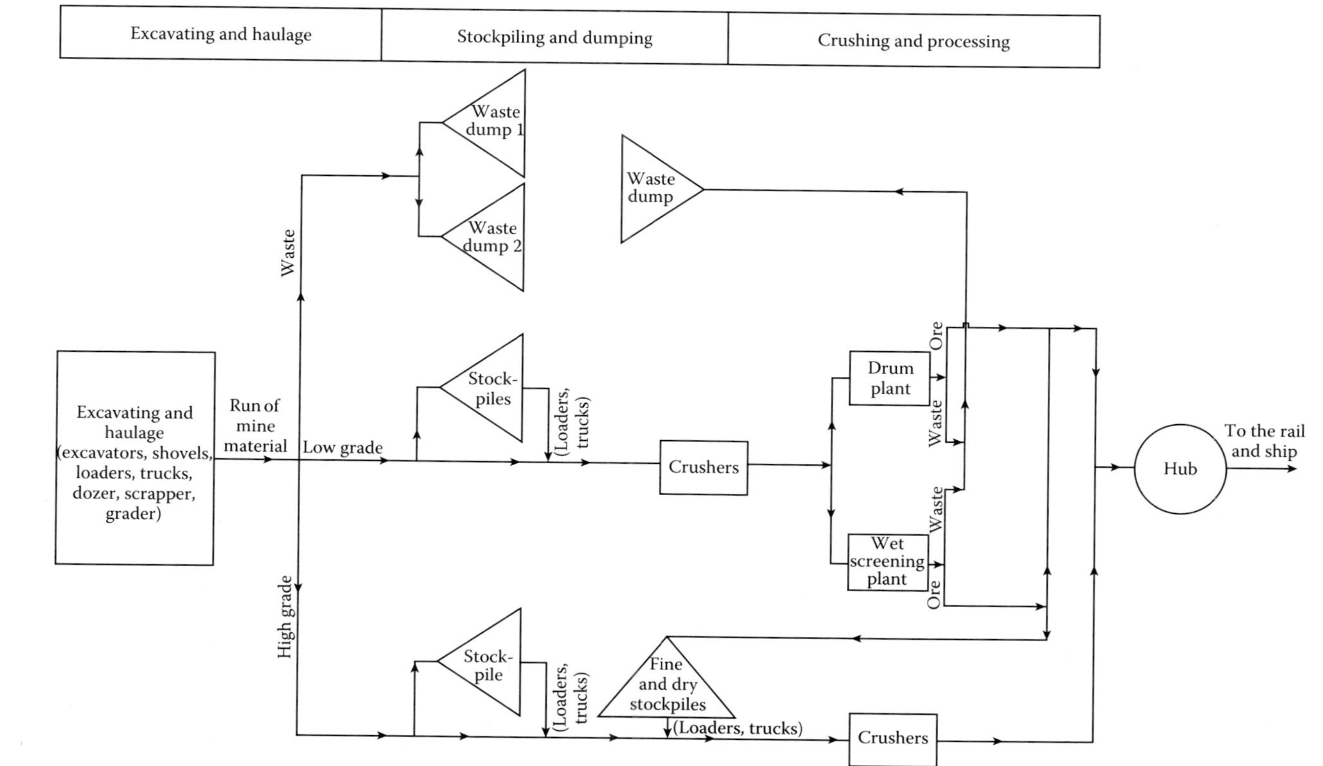

Figure 5.1 A general process flow in an iron ore mine.

5.2.1 Notations

T: number of time periods
t: time period index, $t = 1, 2, \ldots, T$
r: time period index, $r = 1, 2, \ldots, t$
i: block index, $i = 1, 2, \ldots, I$
b_i: tonnage of block i
A: number of attributes
α: attribute (grade) index, $\alpha = 1, 2, \ldots, A$
g_i^α: specification of attribute α in block i
$\Gamma_{p_{i'}}$: set of immediate successors of block i'
u_{id}: unit value of block i when it is sent to destination d
M: number of machines
m: machine index (e.g., excavator, shovel, loader), $m = 1, 2, \ldots, M$
e_{im}: extraction rate of machine m for block i
p_{im}: number of time periods consumed to extract block i by machine m, ($p_{im} = b_i/e_{im}$)
θ_m^t: equal to 1 if machine m is not available in time period t, otherwise 0
P: total number of mills (mineral processing plant)
ρ: mill index, $\rho = 1, 2, \ldots, P$.
W: number of waste dumps
w: waste dump index, $w = 1, 2, \ldots, W$
D: number of destinations ($D = W + P$)
d: destination index, $d = 1, 2, \ldots, D$
$M_\rho^{\min}$: minimum capacity of mill ρ
$M_\rho^{\max}$: maximum capacity of mill ρ
$\varphi_{\alpha d}^{\min}$: minimum required attribute α for destination d
$\varphi_{\alpha d}^{\max}$: maximum required attribute α for destination d

5.2.2 Decision Variables

$$x_{imd}^t = \begin{cases} 1 & \text{if block } i \text{ is being extracted by machine } m \text{ at } t \text{ and sent to destination } d. \\ 0 & \text{otherwise} \end{cases}$$

$y_{ii'}^t$: the binary decision variable to handle precedence if-then constraint

5.2.3 Objective Function

The objective function is to maximize the profit:

$$\sum_{t=1}^{T}\sum_{i=1}^{I}\sum_{m=1}^{M}\sum_{d=1}^{D} x_{imd}^{t} e_{im} u_{id}. \tag{5.1}$$

5.2.4 Constraints

$$\sum_{r=1}^{t-1}\sum_{d=1}^{D} x_{i'md}^{t} - p_{i'm} \geq L\left(y_{ii'}^{t} - 1\right)$$

$$\forall \left\{i, i' \in I \mid i' \neq i, i \in \Gamma_{p_{i'}}\right\}; t = 2,\ldots,T; m = 1\ldots M. \tag{5.2}$$

$$\sum_{m=1}^{M}\sum_{d=1}^{D}\sum_{i=1}^{\Gamma_{p_{i'}}} x_{imd}^{t} \leq L y_{ii'}^{t} \quad \forall \left\{i, i' \in I \mid i' \neq i, i \in \Gamma_{p_{i'}}\right\}; t = 1,2,\ldots,T. \tag{5.3}$$

$$\sum_{t=1}^{T}\sum_{m=1}^{M}\sum_{d=1}^{D} x_{imd}^{t} e_{im} \leq b_i \quad \forall i = 1,2,\ldots,I. \tag{5.4}$$

$$\sum_{i=1}^{I}\sum_{d=1}^{D} x_{imd}^{t} \leq 1 \quad \forall m = 1,2,\ldots,M; t = 1,2,\ldots,T. \tag{5.5}$$

$$\sum_{t=1}^{T}\sum_{i=1}^{I}\sum_{d=1}^{D} x_{imd}^{t} + \sum_{t=1}^{T} \theta_{m}^{t} = T \quad \forall m = 1,2,\ldots,M. \tag{5.6}$$

$$\sum_{m=1}^{M}\sum_{d=1}^{D} x_{imd}^{t} \leq 2 \quad \forall i = 1,2,\ldots,I; t = 1,2,\ldots,T. \tag{5.7}$$

$$\sum_{m=1}^{M}\sum_{i=1}^{I} x_{im\rho}^{t} e_{im} \geq M_{\rho}^{\min} \quad \forall t = 1,2,\ldots,T; \rho = 1,2,\ldots,P. \tag{5.8}$$

$$\sum_{m=1}^{M}\sum_{i=1}^{I} x_{im\rho}^{t} e_{im} \leq M_{\rho}^{\max} \quad \forall t = 1,2,\ldots,T; \rho = 1,2,\ldots,P. \tag{5.9}$$

$$\sum_{m=1}^{M}\sum_{i=1}^{I} x_{imd}^{t}\left(g_i^{\alpha} - \varphi_{\alpha\rho}^{\min}\right) \geq 0$$

$$\forall t = 1,2,\ldots,T;\ \rho = 1,2,\ldots,P;\ \alpha = 1,2,\ldots,\text{A}. \qquad (5.10)$$

$$\sum_{m=1}^{M}\sum_{i=1}^{I} x_{imd}^{t}\left(g_i^{\alpha} - \varphi_{\alpha\rho}^{\max}\right) \leq 0$$

$$\forall t = 1,2,\ldots,T;\ \rho = 1,2,\ldots,P;\ \alpha = 1,2,\ldots,\text{A}. \qquad (5.11)$$

$$x_{imd}^{t} \in \{0,1\}. \qquad (5.12)$$

Equations 5.2 and 5.3 ensure that precedence constraint is satisfied. Precedence constraint indicates that directly related overlying blocks should be mined before extracting the target block. Directly related overlying blocks or precedence relations for the target block are determined by applying slope pattern. There are several slope patterns used to identify precedence relationships. For example, 1:5-pattern, 1:5:9-pattern, and knight's move pattern are commonly used methods to generate precedence relationships (Hochbaum and Chen, 2000). As an example, the 1:5:9-pattern is shown in Figure 5.2. According to this pattern, each block should be connected to five blocks in the upper level and nine blocks in the second upper level. In other words, 14 overlying blocks must be mined before mining the target block. Generally, if the block dimensions or stable slopes vary in different directions or levels, then slope pattern may change. Wright (1990) discussed the details of methods applied to identify precedence relations. Equation 5.4 restricts that at most b_i tones of material can be extracted from block i in the time horizon. Equation 5.5 enforces that each machine (excavator) is working on at most one block in a time period. Equation 5.6 indicates any machine cannot work more than the whole time horizon. In addition, machine may not work in some periods due to the maintenance or other reasons. According to Equation 5.7, a maximum of two machines can work on a block in a time period. Equations 5.8 and 5.9 represent the mill capacity constraints. According to these constraints, the total tonnage of ore material that is sent to the mineral processing plant must not be more that the maximum capacity of the processing plant. In addition, this

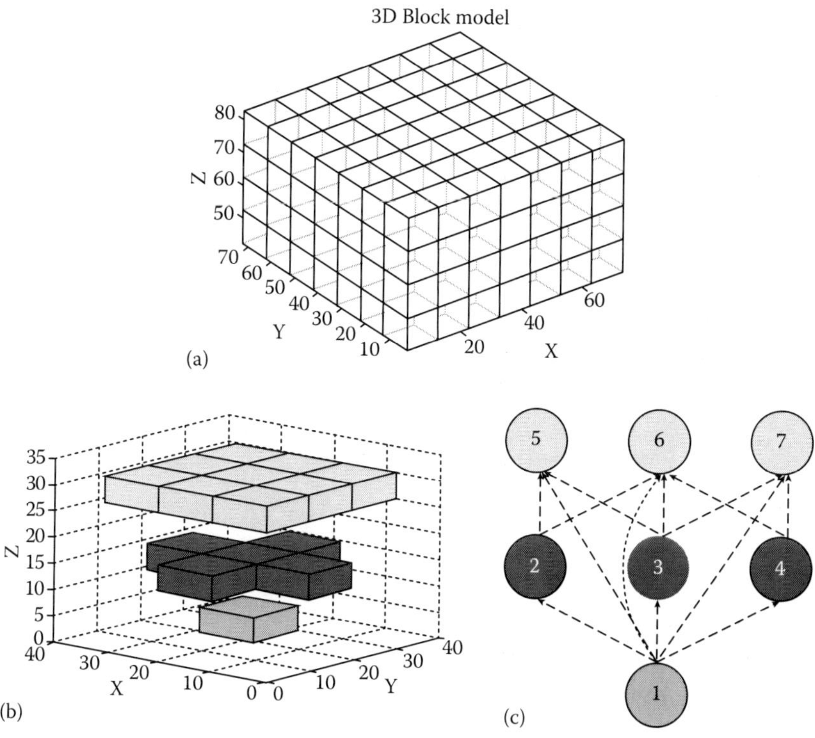

Figure 5.2 (a) A 3D block model; (b) 1:5:9-pattern; to extract a block in level k, 5 blocks in level $k+1$ and 9 blocks in level $k+2$ must be mined first; (c) precedence relations in a side view.

tonnage must not be less than the minimum required feed for the processing plant in time period t. The mill feed must contain required quality of ore content. Equations 5.10 and 5.11 ensure that grade quality for the mill is satisfied. Finally, Equation 5.12 states that decision variables are binary.

5.3 Computational Experiments

To illustrate and validate the proposed model, numerical investigations are accomplished. Several sample data sets are prepared based on the collected data from an iron ore mine in Australia. The characteristics of instances are summarized in Table 5.1.

The proposed model has been coded in the ILOG CPLEX optimizer as a MIP problem and run in a PC Intel Core i7, 2.7 GHz, with 8 GB of RAM, running Windows 7. To optimize a MIP problem,

Table 5.1 Characteristics of Instances

	NUMBER OF			
INSTANCE	BLOCKS	MACHINES	DESTINATIONS	TIME PERIODS (H)
OPBS1	25	4	2	12
OPBS2	25	5	2	12
OPBS3	25	6	2	12
OPBS4	25	4	4	12
OPBS5	25	5	4	12
OPBS6	25	6	4	12
OPBS7	50	4	2	24
OPBS8	50	5	2	24
OPBS9	50	6	2	24
OPBS10	50	4	4	24
OPBS11	50	5	4	24
OPBS12	50	6	4	24
OPBS13	75	4	2	30
OPBS14	75	5	2	30
OPBS15	75	6	2	30
OPBS16	75	4	4	30
OPBS17	75	5	4	30
OPBS18	75	6	4	30
OPBS19	100	4	2	48
OPBS20	100	5	2	48
OPBS21	100	6	2	48
OPBS22	100	4	4	48
OPBS23	100	5	4	48
OPBS24	100	6	4	48
OPBS25	150	4	2	72

CPLEX constructs a tree with the linear relaxation of the original MIP at the root and subproblems at the nodes of the tree. The best node is the node with the best achievable objective function value. The best MIP bound is the best integer objective value among all the remaining subproblem nodes. The relative MIP GAP is the difference between the best integer objective and the best node objective. If CPLEX reaches to the acceptable GAP, branch-and-cut procedure stops and begins polishing a feasible solution. The relative MIP GAP is calculated as follows (Cplex, 2010): $GAP\% = (1 - (|\,Best\,Node\,|/|\,Best\,Integer\,|)) \times 100$. Results of these instances are given in Table 5.2.

In the proposed model, destinations consist of at least one mill and one waste dump; however, the model can be adopted for more mills

Table 5.2 Results of the Numerical Experiments

INSTANCE	NUMBER OF DECISION VARIABLES	NUMBER OF CONSTRAINTS	BEST INTEGER (MILLION $)	MIP SOLVER			CP SOLVER		
				OBJECTIVE VALUE (MILLION $)	GAP%	CPU TIME (S)	OBJECTIVE VALUE (MILLION $)	GAP%	CPU TIME (S)
OPBS1	2,508	645	170.71	170.71	0.0	4	170.71	0.0	420
OPBS2	3,108	660	159.91	159.91	0.0	5	159.91	0.0	610
OPBS3	3,708	673	116.33	116.33	0.0	3	116.33	0.0	840
OPBS4	4,908	695	239.79	239.79	0.0	16	239.79	0.0	180
OPBS5	6,108	709	237.86	237.86	0.0	15	237.86	0.0	320
OPBS6	7,308	723	227.07	227.07	0.0	12	227.07	0.0	1,550
OPBS7	10,080	2,412	345.60	345.44	0.05	10,800	342.16	0.99	10,800
OPBS8	12,480	2,436	325.38	325.38	0.0	42	283.28	12.9	10,800
OPBS9	14,880	2,462	292.91	292.91	0.0	102	20.65	92.9	10,800
OPBS10	19,680	2,508	419.17	419.17	0.0	44	NS	—	10,800
OPBS11	24,480	2,532	397.58	397.58	0.0	54	NS	—	10,800
OPBS12	29,280	2,560	375.97	375.97	0.0	364	NS	—	10,800
OPBS13	19,560	5,694	399.57	399.57	0.0	28	260.05	34.9	10,800
OPBS14	24,060	5,725	350.18	350.[illegible]8	0.0	50	NS	—	10,800
OPBS15	28,561	5,757	NA	NA	—	—	NA	—	—
OPBS16	37,560	5,813	399.57	399.57	0.0		NS	—	10,800
OPBS17	46,560	5,845	372.57	372.57	0.0	37	NS	—	10,800
OPBS18	55,560	5,877	NA	NA	—	—	NA	—	—

(*Continued*)

Table 5.2 (*Continued*) Results of the Numerical Experiments

				MIP SOLVER			CP SOLVER		
INSTANCE	NUMBER OF DECISION VARIABLES	NUMBER OF CONSTRAINTS	BEST INTEGER (MILLION $)	OBJECTIVE VALUE (MILLION $)	GAP%	CPU TIME (S)	OBJECTIVE VALUE (MILLION $)	GAP%	CPU TIME (S)
OPBS19	42,336	13,166	575.09	575.09	0.0	6,300	NS	—	10,800
OPBS20	51,936	13,215	NA	NA	—	—	NA	—	—
OPBS21	61,536	13,264	NA	NA	—	—	NA	—	—
OPBS22	80,736	13,356	575.09	NS	—	10,800	NS	—	10,800
OPBS23	99,936	13,406	736.67	NS	—	10,800	NS	—	10,800
OPBS24	119,136	13,456	NA	NA	—	—	NA	—	—
OPBS25	100,080	38,896	103.60	NS	—	10,800	NS	—	10,800

Note: NA, No feasible solution can be found for this instance; NS, no solution has been found within the predetermined solution time.

and waste dumps for larger mining operation cases. The analysis, as shown in Table 5.2, demonstrates that increasing the number of destinations or machines will not change the complexity of the problem too much. However, the effects of increasing the number of blocks or time periods are more substantial.

The computational experiments show that the OPBS problem with a large number of blocks cannot be solved by MIP solver in a reasonable time and showed that we need a better solution technique to solve this NP-complete problem. The input data, especially for block information, are changed and updated frequently; therefore, the solution algorithm must be quick enough to solve the problem in reasonable time by a standard computer.

The infeasibility of instances OPBS15, OPBS18, OPBS20, OPBS21, and OPBS24 come from the input data, where there is not enough qualified material for the mill.

In this chapter, the ability of constraint programming (CP) to solve the OPBS problem is also investigated. In CP, a solution is a feasible set of the variables that fulfill the constraints (Van Beek and Chen, 1999). In this study, CPLEX CP engine has been used to solve the instances. The results of CP are presented in Table 5.2. Results state that high-quality solutions are obtained by CP for small-size instances of this problem. However, computational results of larger instances indicate that CP is not a timely efficient approach to solve this problem. The reason for this observation may arise from the fact that there are many binary variables that are nominated to be assigned a value (0 or 1). Therefore, during the process of CP, a large number of branching steps must be carried out that lead to increase the solution time. An alternative treatment approach to overcome this drawback can be combining the CP with the constructive heuristic method in order to direct the CP branching steps by finding an initial feasible solution. Investigation and verification of this alternative are left to the future work.

5.4 Conclusion

In summary, we have developed a MIP model for the OPBS problem. In the proposed model, real-life constraints including precedence relationships, mill capacity, resource capacity, and grade control are considered. The applicability of the model has been tested by performing computational experiments on the real-base data. Numerical investigations demonstrate that the OPBS problem is a complex problem that cannot be solved by exact solution techniques in a reasonable time. In addition, the capability of CP approach to solve this problem has been investigated. The computational experiments of CP by the CPLEX CP engines state that CP may not be a timely efficient approach for this problem. However, to prove the latest claim, more studies should be conducted in the future research. In addition, a state-of-the-art meta-heuristic algorithm will be developed to solve large-size instances.

Acknowledgment

The authors acknowledge the support of CRC ORE, established and supported by the Australian Government's Cooperative Research Centers Program.

References

Asad, M. W. A. and Topal, E. (2011). Net present value maximization model for optimum cut-off grade policy of open pit mining operations. *The Journal of The Southern African Institute of Mining and Metallurgy, 111*, 741–750.

Askari-Nasab, H., Awuah-Offei, K., and Eivazy, H. (2010). Large-scale open pit production scheduling using mixed integer linear programming. *International Journal of Mining and Mineral Engineering, 2*(3), 185–214.

Bienstock, D. and Zuckerberg, M. (2010). Solving LP relaxations of large-scale precedence constrained problems. In F. Eisenbrand and B. Shepherd (eds.), *Integer Programming and Combinatorial Optimization* (pp. 1–14). Springer, Berlin, Germany.

Boland, N., Dumitrescu, I., Froyland, G., and Gleixner, A. M. (2007). LP-based disaggregation approaches to solving the open pit mining production scheduling problem with block processing selectivity. *Computers & Operations Research, 36*, 1064–1089.

Caccetta, L. and Hill, S. P. (2003). An application of branch and cut to open pit mine scheduling. *Journal of Global Optimization, 27*, 349–365.

Chicoisne, R., Espinoza, D., Goycoolea, M., Moreno, E., and Rubio, E. (2012). A new algorithm for the open-pit mine production scheduling problem. *Operations Research, 60*(3), 517–528.

Cplex, I. I. (2010). *12.2 User's Manual.* IBM, United States.

Cullenbine, C., Wood, R. K., and Newman, A. (2011). A sliding time window heuristic for open pit mine block sequencing. *Optimization Letters, 5*, 365–377.

Dagdelen, K. and Johnson, T. B. (1986). Optimum open pit mine production scheduling by Lagrangian parameterization. In *Proceedings of the 19th Symposium of APCOM*, Jostens Publications, State College, PA (pp. 127–141).

Denby, B. and Schofield, D. (1994). Open-pit design and scheduling by use of genetic algorithms. *Transactions of the Institution of Mining and Metallurgy. Section A. Mining Industry, 103*, A21–A26.

Ferland, J. A., Amaya, J., and Djuimo, M. S. (2007). Application of a particle swarm algorithm to the capacitated open pit mining problem. *Studies in Computational Intelligence (SCI), 76*, 127–133.

Gershon, M. (1987). Heuristic approaches for mine planning and production scheduling. *International Journal of Mining and Geological Engineering, 5*, 1–13.

Hochbaum, D. S. and Chen, A. (2000). Performance analysis and best implementations of old and new algorithms for the open-pit mining problem. *Operation research, 48*(6), 894–914.

Johnson, P. V., Evatt, G., Duck, P., and Howell, S. (2011). The determination of a dynamic cut-off grade for the mining industry. In S. I. Ao and L. Gelman (eds.), *Electrical Engineering and Applied Computing* (Vol. 90, Chapter 32, pp. 391–403). Springer-Verlag, Berlin, Germany.

Kozan, E. and Liu, S. Q. (2011). Operations research for mining: A classification and literature review. *ASOR Bulletin, 30*(1), 2–23.

Kumral, M. and Dowd, P. A. (2005). A simulated annealing approach to mine production scheduling. *The Journal of the Operational Research Society, 56*(8), 922–930.

Lamghari, A. and Dimitrakopoulos, R. (2012). A diversified Tabu search approach for the open-pit mine production scheduling problem with metal uncertainty. *European Journal of Operational Research, 222*(3), 642–652.

Lark, R. M., Cullis, B. R., and Welham, S. J. (2006). On spatial prediction of soil properties in the presence of a spatial trend: The empirical best linear unbiased predictor (E-BLUP) with REML. *European Journal of Soil Science, 57*, 787–799.

Menabde, M., Froyland, G., Stone, P., and Yeates, G. (2004). Mining schedule optimisation for conditionally simulated orebodies. In *Proceedings of the International Symposium on Orebody Modelling and Strategic Mine Planning: Uncertainty and Risk Management*, Perth, Western Australia (pp. 347–352).

Myburgh, C. and Deb, K. (2010). Evolutionary algorithms in large-scale open pit mine scheduling. In *Proceedings of the 12th Annual Conference on Genetic and Evolutionary Computation* (pp. 1155–1162). ACM, New York.

Newman, A. M., Rubio, E., Caro, R., and Eurek, K. (2010). A review of operations research in mine planning. *Interfaces, 40*(3), 222–245.

Osanloo, M., Gholamnejad, J., and Karimi, B. (2008). Long-term open pit mine production planning: A review of models and algorithms. *International Journal of Mining, Reclamation and Environment, 22*(1), 3–35.

Ramazan, S. (2007). The new fundamental tree algorithm for production scheduling of open pit mines. *European Journal of Operational Research, 177*, 1153–1166.

Ramazan, S. and Dimitrakopoulos, R. (2004). Recent applications of operations research and efficient MIP formulations in open pit mining. *Mining, Metallurgy, and Exploration Transactions 316*, 73–78.

Sattarvand, J. (2009). Long term open pit planning by ant colony optimization. PhD Dissertation, RWTH Aachen University, Aachen, Germany, p. 144.

Sattarvand, J. and Niemann-Delius, C. (2008). Perspective of metaheuristic optimization methods in open pit production planning. *Gospodarka Surowcami Mineralnymi, 24*(4), 143–156.

Tolwinski, B. and Underwood, R. (1996). A scheduling algorithm for open pit mines. *IMA Journal of Mathematics Applied in Business & Industry, 7*, 247–270.

Van Beek, P. and Chen, X. (1999). CPlan: A constraint programming approach to planning. In *AAAI/IAAI*, Orlando, FL (pp. 585–590).

Vann, J., Bertoli, O., and Jackson, S. (2002). An overview of geostatistical simulation for quantifying risk. In *Proceedings of Geostatistical Association of Australasia Symposium "Quantifying Risk and Error"*, Perth, Western Australia (p. 1).

Verly, G. (2005). Grade control classification of ore and waste: A critical review of estimation and simulation based procedures. *Mathematical Geology, 37*(5), 451–475.

Wright, A. (1990). *Open Pit Mine Design Model: Introduction with FORTRAN 77 Programs.* Trans Tech Publication, Clausthal-Zellerfeld, Germany.

6

Locating Temporary Storage Sites for Managing Disaster Waste, Using Multiobjective Optimization

KIVANÇ ONAN, FÜSUN ÜLENGIN, AND BAHAR SENNAROĞLU

Contents

6.1 Introduction

The disaster wastes (DWs) may include recyclable, reusable, and hazardous materials. For instance, asbestos is a hazardous waste, which may be found in disaster debris and cause serious illnesses when not properly disposed. Because of the risks in debris composition, separation and treatment of materials is a serious problem. The composition of waste is also an opportunity in terms of environmental sustainability.

Generally, collected DW is disposed at landfill areas all around world. But uncontrolled disposal of the waste is environmentally and economically harmful.

There are ways of proper disaster waste management (DWM). The ingredients of DW is similar to regular construction and

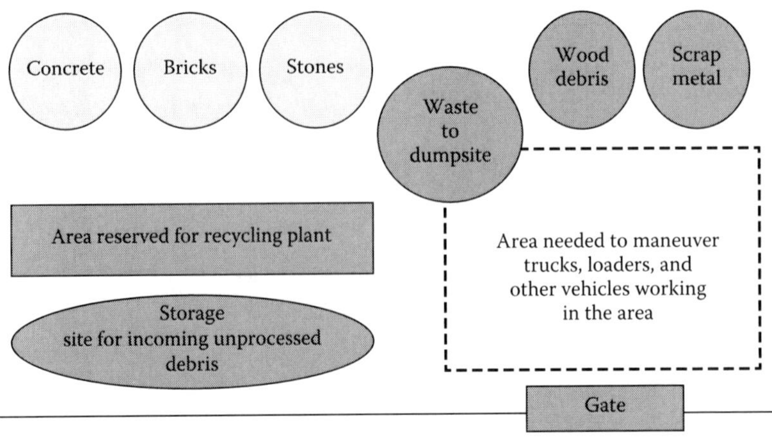

Figure 6.1 Suggested layout of a temporary storage site. (From Söder, A.B. and Müller, R., *Disaster Waste Management Guidelines*, United Nations Office for the Coordination of Humanitarian Affairs Environmental Emergencies Section [UNEP/OCHA Environment Unit], Geneva, Switzerland, 2011.)

demolition (C&D) wastes. So the C&D waste treatment methods can be a starting point when DW is considered. The most common way applied to manage the C&D waste is collecting all waste and transporting to dumpsites for final disposal without any treatment, just like the present applications of DWM. One other way is to transport collected waste to a recycling facility and separate reusable and recyclable parts from the debris either before or after transporting them. Another way, which is the most environmentally effective one, is to separate all materials at a temporary site (Figure 6.1), so that the landfill disposal of untreated hazardous wastes is minimized and all applicable materials can be recycled or reused (Söder and Müller, 2011).

Separated materials can be immediately reusable for a new construction, which may be important for the immediate reconstruction activities after a disaster happened, or can be recycled for also immediate or future use in these reconstruction activities. So this last way of treating waste can be said to be a more proper way, both considering environment and economics, for DWM.

A special needs research session on DWM was held in 2010 called *Intercontinental Landfill Research Symposium* in Japan, and one of the three main areas emphasized as future research areas was emergency

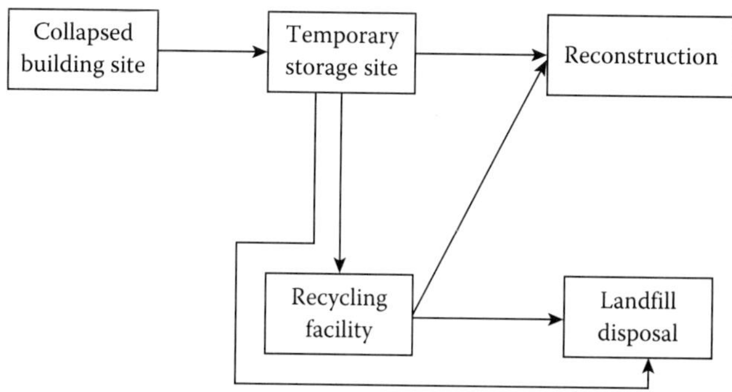

Figure 6.2 Concept of temporary storage site.

temporary storage areas (Milke, 2011). Since the emergency response circumstances cause poor storage area planning for DW, there is a need for research on planning temporary storage areas for keeping, separating, and recycling the DW (Milke, 2011).

Separating hazardous materials from DW is an important part of the process, which must be carried out with extreme care. That is especially because of asbestos and other hazardous and harmful materials, which may be found in the composition of DW. The concept of temporary storage facilities (Figure 6.2) is therefore very important in order to properly treat these hazardous materials in the debris for decreasing the risks to people's health. As an example to the harms of hazardous materials, the major risk caused by asbestos can be given: once the building parts that are made of asbestos are broken, during a demolition, loading, transportation, treatment of a building, or disposing DW, huge amounts of very small particles spread through air, and breathing or contacting these particles may cause several diseases; even there is a high risk of cancer (Söder and Müller, 2011).

In order to manage DW properly, first step is assessing the risk and then estimating the disaster-related damages or in other words loss estimation. After a survey on disaster risk assessment and loss estimation studies, a tool that was developed by the researchers at Kandilli Observatory and Earthquake Research Institute (KOERI), which is called ELER, is selected.

ELER is a tool that is developed to rapidly estimate the loss after the earthquakes with a potential of damage. This is very vital for efficient emergency response and also for informing the community. ELER is MATLAB®-based tool, which is used to create an earthquake and assess the building damage and casualties according to this simulated earthquake (Hancılar et al., 2010). Software includes a module called Hazard, which simulates an earthquake according to given parameters and three modules for loss estimation. These three modules are named as level 0, level 1, and level 2. Hazard module simulates the earthquake and produces shake maps according to the given parameters of ground motion. The parameters are peak ground acceleration, peak ground velocity, and spectral acceleration, and also depth, magnitude, broken fault, etc., for simulating an earthquake. These parameters are used to calculate ground motion prediction equations (Hancılar et al., 2010).

ELER software has three levels for hazard assessment. The estimated ground motion parameters, which are calculated with hazard module, are used to produce loss estimations according to the level of assessment. Demographic data and the given building inventory are used to calculate the loss estimations. The building inventory includes the number and properties of buildings and also the number of dwellings in each building. These values are used to transform the number of damaged buildings to amount of waste.

There are three parameters, which are used for calculating damage of buildings and casualties. These parameters are population distribution, building inventory data, and vulnerability relationships. Level 1 module produces the estimations of building damages and also casualty distributions. This module is used to produce damage estimations for the model proposed in this study.

The aim of this research is to determine the candidate locations of temporary storing areas of DW while minimizing the total cost, and at the same time to minimize the amount of population subject to risk.

To this point, main motives and first steps of the study are explained in detail; the background, methodology, model, solution, and discussions are given in the following sections.

6.2 Background of the Study

This section is constructed to understand the underlying ideas and motives for the proposed framework. For this purpose, the related literature was reviewed. First, the two main reports about Istanbul and the expected earthquake were reviewed. These are Earthquake Master Plan of Istanbul and JICA Report about Istanbul Earthquake. These reports include the most detailed researches about two alternative estimated earthquake models (Model A and Model C) for Istanbul. These reports also include brief information on estimations about the ratio of heavily damaged buildings for each district of Istanbul (Ansal et al., 2003; Ikenishi et al., 2002).

Another field to be reviewed is DWM. There are four main issues when treating DW: collection of waste, reuse, recycling, and landfill disposal. When planning collection of waste, hazardous materials, which the debris may contain, must be considered in order to keep the process environmentally sustainable for people's health. The most effective way of dealing with hazardous materials in a sustainable way is to minimize the landfill disposal.

As it was mentioned before, the aim of this research is to determine the candidate locations for locating the temporary storage areas while minimizing the total cost, and at the same time to minimize the amount of population subject to risk related to storing, transportation, and recycling of DW.

In order to optimize these objectives, the problem must be modeled and solved mathematically. Since the problem is of multiobjective type (minimization of total cost and minimization of risk on population), related studies were reviewed. Multiobjective evolutionary optimization methods are state of art, so these methods are reviewed in detail. (Collette and Siarry, 2003; Deb, 2001, 2008; Marler and Arora, 2004) Brief information about DWM studies will be given in the following section.

6.2.1 Disaster Waste Management

Disasters can cause large amounts of waste. DW is mainly caused by the disaster itself, but waste may occur also during recovery stage. These wastes can cause health risks in case of contact with hazardous wastes, such as asbestos, pesticides, oils, and solvents. Also DWs may block relief efforts (Brown et al., 2011; FEMA, 2010; Milke, 2011; Söder and Müller, 2011).

Having so many negative impacts on health and relief efforts, DW may also contain materials, which may be needed for recovery, such as concrete, steel, and wood. These valuable materials can be used for reconstruction of the affected area and so decreases the need for natural resources.

Because of the earlier-mentioned issues, managing DW is a critical issue. Proper management reduces the risks and provides recovery opportunities.

Recent studies show that this importance is not recognized enough that in most cases, DWM means collecting and dumping waste without control, separation, and recycling or reuse. This kind of treatment to waste may cause environmental problems, which may affect public health and may result with contamination of valuable lands, and further actions to be taken to deal with this problem will cause extra costs.

There was a need for a guideline for the appropriate treatment of DWM. Especially United Nations and some researchers made recent studies on DWM (Brown et al., 2011; Milke, 2011).

The common objectives of these studies are minimizing risk to health, minimizing risk to environment, and ensuring recycling or reuse for the benefit of the affected people. Also, another objective being considered by the authorities is the cost of these operations.

DW issues are as follows: uncollected waste from damaged buildings, dumping in unsuitable sites, problems in solid waste services due to disaster, and uncontrolled disposal of hazardous and infectious waste.

These issues cause some impacts as follows: damaged sites can be considered as dumping areas, which increases the amount of waste to be collected, effect of closely located disposal sites on health, destructing usable lands, effects on water resources, additional cost of responding to these impacts, health risks occurred during inhalation or exposure.

Since the main focus of this study is earthquake disaster, a deeper focus is needed. Earthquake damages may make it difficult to separate hazardous wastes from other wastes since a total collapse may happen. Recovery requires heavy machinery, which increases the cost and difficulty to collect wastes. But the most important issue is that the quantity generated by an earthquake may be very high compared to other disaster types.

United Nations Environment Programme's (UNEP) DWM guidelines suggest preparing a contingency plan prior to disaster, and also it is suggested that this plan should be cost-effective and should increase control over waste management (Söder and Müller, 2011).

UNEP guidelines give a pathway to deal with DW, and the steps of this pathway include the following:

- Forecasting amounts of waste and debris
- Monitoring current capacity for waste and debris management
- Selecting waste and debris storage sites prior to disaster
- Creating a debris removal strategy

Researchers indicate that increasing number and intension of natural disasters increases the importance of efficient and low impact recovery (Brown et al., 2011). Writers also indicate that importance of DWM has been recognized since "Planning for Disaster Debris" was published by United States Environmental Protection Agency in 1995. This report was also updated in 2008.

The earlier-mentioned literature review highlighted that the importance of DWM is almost not realized by the researchers from a sustainability perspective. DWM-related studies mostly focus on guiding authorities on how to make a plan prior to the disaster and how to act after the disaster (FEMA, 2010; Söder and Müller, 2011). But these studies do not include mathematical models optimizing both environmental effect and cost. So it can easily be seen that both multiobjective approaches, for sustainable and economic management, and DW-related studies are needed to be done by the researchers (Milke, 2011).

6.3 Methodology

Figure 6.3 represents the general framework of the methodology proposed in this study. Shortly, the study consists of three steps: producing loss estimations with ELER, obtaining data produced by ELER and producing OD matrix with ArcGIS, and finally using these data to determine the candidate points with a multiobjective optimization (MOO) model. The solution gives the Pareto-optimal set or front. Final step is to choose solutions from the set according to the preferences of decision maker (DM), which is called higher-level information. As an example to higher-level information, total distance to landfill disposal sites or distance to water supplies may be given.

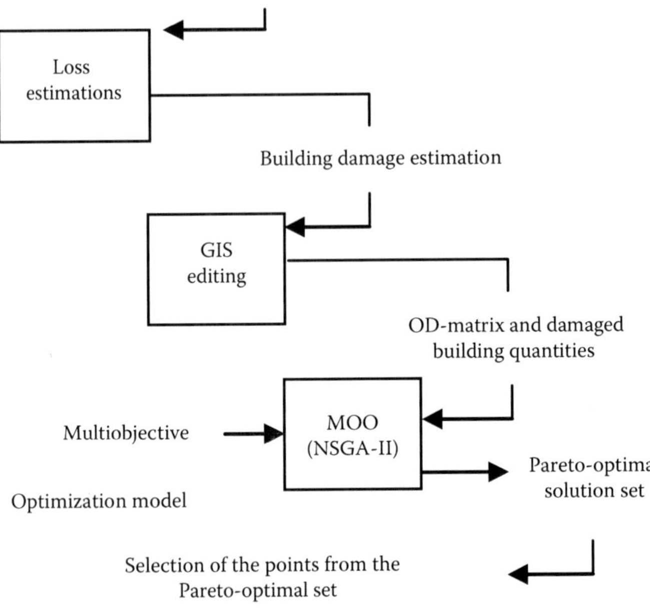

Figure 6.3 General framework of the study.

In this case, since the aim is to determine the candidate locations for temporary storage facilities, all points selected will be considered.

The formulation of the mathematical model is given as follows. This mathematical model is a biobjective model for deciding candidate locations and afterward selection of the most preferred solution from the Pareto-optimal front:

$$\min z = \sum_{i=1}^{n} x_i A_i \tag{6.1}$$

$$\min z = \sum_{i=1}^{n} x_i P_i \tag{6.2}$$

subject to

$$\sum_{i=1}^{n} x_i \geq M \tag{6.3}$$

where x_i is the binary integer variable; it is equal to 1 if a temporary storage site is located to *i*th cell. A_i is the average weighted distance of *i*th cell from other cells. So Equation 6.1 represents the total average weighted distance of temporary storage sites to waste source points. P_i is the population of *i*th cell, and Equation 6.2 represents the total population in the cells, which contains a temporary storage site. This equation minimizes the amount of population subject to risks of recycling and separation facilities. M is the minimum number of candidate locations to be selected to let the model locate at least one site, and this number can be determined by DM.

This mathematical model is solved by using nondominated sorting genetic algorithm (NSGA-II) (Deb, 2011). This is an elitist evolutionary genetic algorithm (GA)–based methodology, which is called NSGA-II. Here the term called domination should be explained. Let us consider a biobjective (minimization type) problem with equal importance of these functions. A pair of solutions is said to be nondominated if none of them can be marked as a better one comparing both of the objective function values. The details related to the applied algorithm, including parameters of GA, are given in Section 6.4.

Other issues that should be explained about the applied methodology are the usage of GIS and multiobjective optimization methods. GISs are being used to create and analyze geographical systems. ELER software produces outputs in shapefile format, and these files can only be opened and edited by GIS software. GIS can also be used to create origin-destination (OD) cost or distance matrices. ArcGIS is the most commonly used GIS software. All disaster data created by ELER can be edited and viewed via ArcGIS. Also, ArcGIS has tools stored under network analyst toolbox. These tools include several network analyses used for solving problems such as routing or allocation. Origin destination matrix can be calculated by defining source and demand points. The road network is needed to calculate distances. All shapefile formatted files can be edited by ArcGIS. The integration of loss estimation tools with GIS tools is very crucial for the evaluation of the results and editing of these results for further researches.

The proposed solution methodology is an elitist multiobjective evolutionary optimization method. NSGA-II, SPEA, and PAES are the latest elitist methodologies presented to solve multiobjective models in the most efficient way. The NSGA-II is considered as one of the most robust methods for solving such problems. ELER's building damage values and ArcGIS OD matrix values are used to build multiobjective model to minimize both total cost and amount the distance of storage sites to people.

6.4 Results

The evolutionary multiobjective solution procedure was applied to the problem, and all possible candidate locations were determined. These are the points that take place in at least one solution in the Pareto-optimal front. Figure 6.4 represents all points considered in calculating candidate points. The total number of the cells in the grid is 193, covering all European side of Istanbul (west of Bosporus).

The structure of the studied map is grid based. A grid is composed of cells, which are represented as points in the map. These cells are $0.05° \times 0.05°$, and center points are represented. These are all points in the solution space where a temporary storage site can be located. Each cell includes the information of population and building inventory in that cell.

Figure 6.5 represents the selected 26 points as a result of solution procedure. These points are demonstrated with light-colored circles. As can be seen from the figure, the selected cells mostly located to cells, which both close to city center and also less populated compared to the cells in the city center.

The proposed model is solved using proposed solution methodology, which was explained in the previous section. Since the solution methodology is a GA-based evolutionary algorithm and it searches for better solutions, it requires that parameters of algorithm's operations should be determined by having several runs or trials. For instance, one of the important parameters to be determined is the number of total generations to be created, or, in other words, the termination criterion, which is 10,000 iterations in this case. All the other parameters of the algorithm were fixed (crossover rate = 0.7, mutation probability = 0.0005).

For the final run of the actual model, the parameters can be summarized as follows. Shortly, the mathematical model is composed of

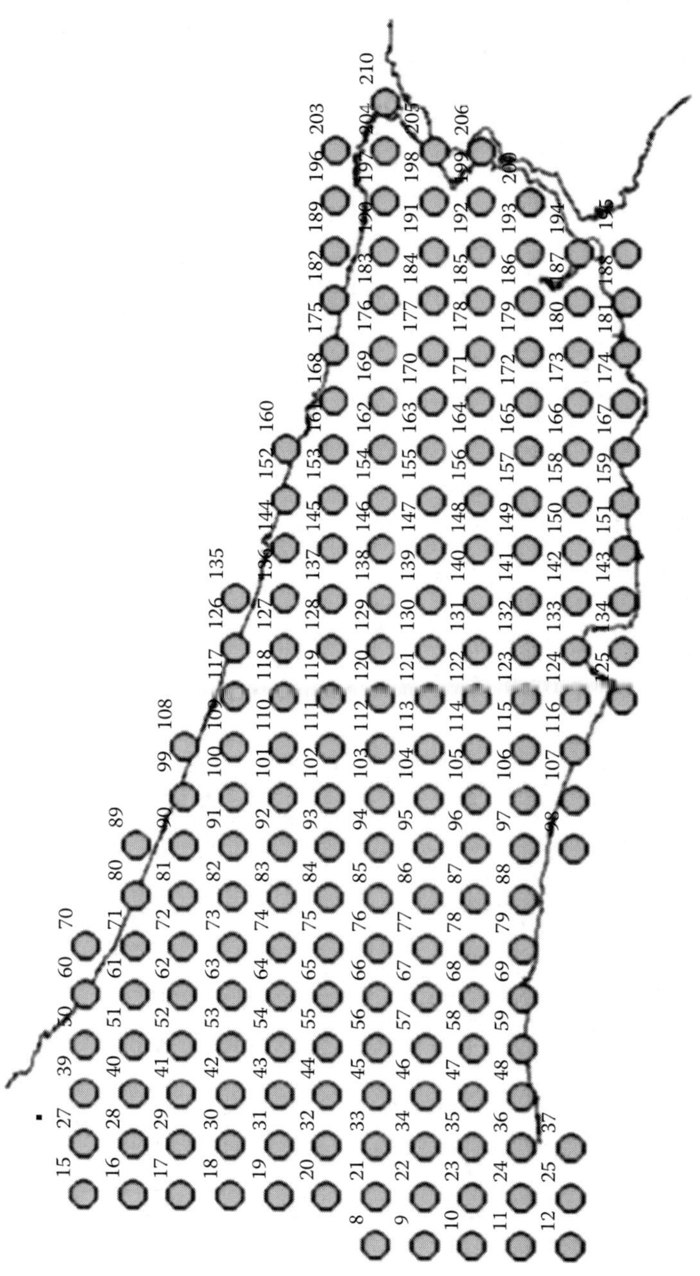

Figure 6.4 The 0.05×0.05 cell grid of Istanbul (points represent the center of each cell for west of Bosporus).

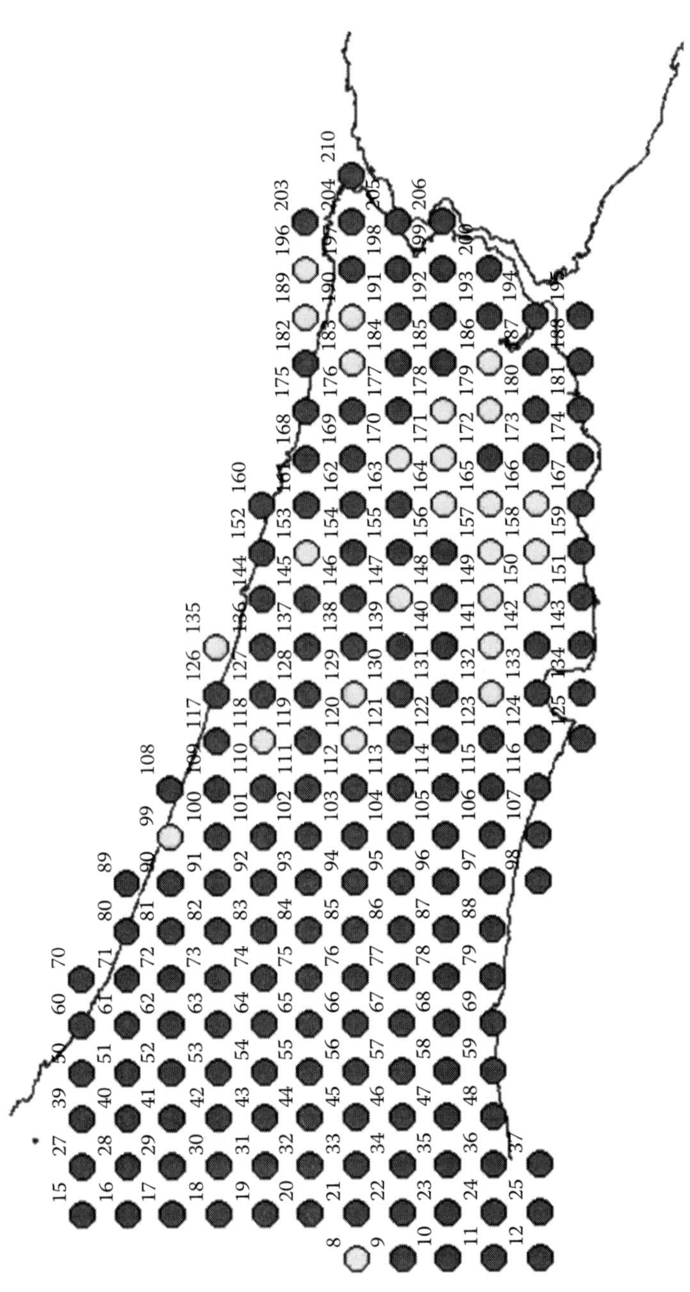

Figure 6.5 Map representing all solutions of first stage (all candidate locations).

two objective functions (distance and population), and then evolutionary multiobjective optimization approach is used for solving this model. This approach is based on GA, where the crossover operation is single-point type and mutation operation is bitwise type.

6.5 Discussion

This study aims to propose a solution methodology for DWM in a sustainable way. In this context, the west side of Bosporus of Istanbul was chosen as a case study. First, the risk of disaster in Istanbul is emphasized with related background. Next, a suitable disaster hazard assessment methodology is searched. Studies of Turkey's principal earthquake institute (KOERI) were surveyed. The studies published by the researchers in KOERI and from all around world were reviewed. A tool called ELER is decided to be used for risk assessment and disaster loss estimation in order to obtain the data needed to calculate waste amount estimations.

One of the scenarios of expected earthquake, which was highly expected to happen, is used to create a hazard simulation in ELER. This scenario is an earthquake with a magnitude of 7.5 in the Richter scale. Then by using the building database and loss estimation modules, the amount of damaged buildings was calculated in each damage category. These numbers were used to calculate the amount of DW in each cell of the GIS based grid, used by ELER for assessing hazards.

A research made about 1999 Marmara earthquake supplied very important information in transforming the number of damaged buildings into the amount of waste generated in tons. Baycan's simple formula is used to transform the number of buildings to approximate waste amount (Baycan, 2004).

Damage data are transformed in tons, then a framework is constructed for solving the candidate temporary storage sites location problem, which includes locating temporary storage sites to cells. This framework has a solution procedure to determine candidate locations.

In terms of sustainability, minimization of distance covered is not satisfactory. Another objective was defined for the model. Since the transportation, disposal, and recycling operations are hazardous and harmful, it is important to keep these sites as far as possible from population. To achieve this goal, another objective is included into the model: minimization of total population in selected cells for locating temporary sites. The framework turned into a biobjective location problem.

For solving biobjective optimization problem, multiobjective optimization solution methods were reviewed. These methods were classified differently in several surveys. In general, these methods can be classified as a priori, a posteriori, interactive, and evolutionary methods.

The review of multiobjective optimization methods showed that for large-scale problems with conflicting objectives, evolutionary methods usually promise a better performance. So focus on multiobjective optimization methods was narrowed to evolutionary methods for this study.

Evolutionary multiobjective optimization methods are GA-based search algorithms. These methods can be classified as nonelitist and elitist methods, where elitist methods have an elite solution preserving mechanism. Since elitist methods were evolved from classic evolutionary methods, they are more up to date, and they promise better performance according to the previous studies.

Elitist evolutionary multiobjective optimization methods were reviewed to select an elitism mechanism for solving the two-stage, biobjective location-allocation optimization problem. This research showed that most of the procedures have a similar framework to NSGA-II solution methodology, which was developed by Kalyanmoy Deb (2011).

After selecting an appropriate solution method, the problem is formulated as a biobjective elitist evolutionary location optimization model. This model is solved using evolutionary and elitist NSGA-II solution methodologies. And results were represented in the previous section.

The result of the problem is a set of solutions, which is named as Pareto-optimal front. These Pareto-optimal solutions need a final discussion to select one or more solutions from the set. This final discussion requires higher-level knowledge. Higher-level knowledge is any information, which can be used to favor a solution or solutions from the Pareto-optimal solutions set. In this case, the points with any appearance in all set are discussed to be the final solution, where it also provides to manage the DW with the widest range of the number of temporary storage facilities selection.

6.6 Conclusion

The most important contribution of this study is the proposed framework, which integrates risk assessment with multiobjective modeling for planning DWM. This integration is provided by incorporating several methods and tools: geographical information systems, earthquake loss estimation, and evolutionary multiobjective optimization. This framework is a new approach for sustainable DWM. Since DWM and MOO are emerging fields, new methods can be applied to similar problems in the context of this framework in future. Also, the future loss estimation methods can be discussed in this way.

This study aimed to determine the candidate locations of temporary storage sites for DWM. But the framework can be used for different problems, which may occur after a disaster. For example, by using suitable loss estimation tools, infrastructure damages can be estimated, and considering several objectives, such as cost and priority, a postdisaster repairing plan can be produced.

The proposed framework is applied to a postdisaster scenario, but predisaster planning studies can also be done within this context. For example, the prioritized areas for urban transformation can be determined by using a similar approach.

The information and regulations about temporary storage sites are technically insufficient or even do not exist so there is a need for studies about the design of these facilities. The mathematical model presented in this study is not including constraints based on the technical limitations of a temporary storage site, so in future researches, such kind of constraints can be added to the model.

The transportation is considered as covered distances in this model. This objective can be calculated as cost of transportation. Also instead of calculating the costs using constant values for carrying waste, cost can be calculated dynamically by changing the value of the cost dynamically depending on the distance traveled.

References

Ansal, A., Özaydın, K. et al. (2003). *Earthquake Master Plan of Istanbul.* Istanbul, Turkey: Istanbul Metropolitan Municipality.

Baycan, F. (2004). Emergency planning for disaster waste: A proposal based on the experience of the Marmara earthquake in Turkey. In *2004 International Conference and Student Competition on Post-Disaster Reconstruction "Planning for Reconstruction"*, Coventry, U.K.

Brown, C., Milke, M., and Seville, E. (2011). Disaster waste management: A review article. *Waste Management*, 31, 1085–1098.

Collette, Y. and Siarry, P. (2003). *Multiobjective Optimization Principles and Case Studies.* Berlin, Germany: Springer.

Deb, K. (2001). *Multi-Objective Optimization Using Evolutionary Algorithms* (1st edn.). Wiltshire, U.K.: John Wiley & Sons, Ltd.

Deb, K. (2008). Introduction to evolutionary multiobjective optimization. In J. Branke, K. Deb, K. Miettinen, and R. Slowinski (Eds.), *Multiobjective Optimization* (1st edn., pp. 59–96). Heidelberg, Germany: Springer-Verlag.

Deb, K. (2011). *Multi-Objective Optimization Using Evolutionary Algorithms: An Introduction*. Kanpur, India: Indian Institute of Technology.

FEMA. (2010). *Debris Estimating Field Guide.* U.S. Department of Homeland Security, http://www.fema.gov/pdf/government/grant/pa/fema_329_debris_estimating.pdf.

Hancılar, U., Tüzün, C., Yenidoğan, C., and Erdik, M. (2010). ELER software—A new tool for urban earthquake loss assessment. *Natural Hazards and Earth System Sciences*, 10(12), 2677–2696.

Ikenishi, N., Kadota, T. et al. (2002). The study on a disaster prevention/mitigation basic plan in Istanbul including microzonation in the Republic of Turkey. Istanbul, Turkey: Japan International Cooperation Agency.

Marler, R. and Arora, J. (2004). Survey of multi-objective optimization methods for engineering. *Structural and Multidisciplinary Optimization*, 26(6), 369–395.

Milke, M. (2011). Disaster waste management research needs. *Waste Management*, 31, 1.

Söder, A. B. and Müller, R. (2011). *Disaster Waste Management Guidelines.* Geneva, Switzerland: United Nations Office for the Coordination of Humanitarian Affairs Environmental Emergencies Section (UNEP/OCHA Environment Unit).

7

Using Earthquake Risk Data to Assign Cities to Disaster-Response Facilities in Turkey

AYŞENUR SAHIN, MUSTAFA ALP ERTEM, AND EMEL EMUR

Contents

7.1 Introduction

Turkey is located at one of the most active earthquake regions of the world. It is the third in the world in terms of human loss and eighth in terms of the number of people affected by an earthquake (AFAD 2012). The only unchanging reality of Turkey besides the political events and the changes of economic conditions that took place during the years is *the earthquake*.

Most of Turkey's population can be considered as risky because of the North Anatolian Fault (NAF) line. Several earthquakes have been reported in this geographical region. In August 17, 1999, Marmara earthquake took place on the western part of NAF line with a magnitude of 7.4 on the Richter scale (Görmez et al. 2011). This major earthquake marks a turning point in the field of disaster management and coordination of disaster relief activities in Turkey. This earthquake, which caused a great loss of life and property, has revealed that the issue of disaster management in Turkey needed to be reconsidered (AFAD 2012).

There is uncertainty in the nature of disasters (e.g., earthquakes) because the timing and location them cannot be predicted beforehand. This uncertainty affects the proper management of disaster relief operations. It has been observed that in different locations of Turkey, earthquakes show different destruction powers. The severity of the earthquake and building quality might be considered as the main source of this difference. On the other hand, when a particular fault line is taken into account, it can be inevitably seen that some locations in Turkey have a higher risk of experiencing devastating earthquakes than the others. In our study, we defined this potential as "the earthquake risk."

Humanitarian logistics is defined as "the process of planning, implementing and controlling the efficient, cost-effective flow and storage of goods and materials, as well as related information, from the point of origin to the point of consumption for the purpose of alleviating the suffering of vulnerable people" (Thomas and Kopczak 2005). Activities in humanitarian logistics include preparedness, planning, procurement, transport, warehousing, tracking and tracing, and customs clearance (Thomas and Kopczak 2005). Similar to the business supply chain and logistics activities, humanitarian logistics includes diverse activities like procurement and prepositioning. Before the onset of disaster, the relief items are procured from global or local sources and stored in the warehouses. Therefore, prepositioning provides time and place utility since the time and location of the disasters cannot be predicted beforehand. Moreover, after the disaster onset the warehouses are continuously supplied with amenities from the suppliers because of the flow of relief items from warehouses to disaster locations. Therefore, planning the storage locations of relief supplies and selecting these locations in terms of vulnerability is a crucial job before disasters for humanitarian relief organizations. This study aims to assign demand points to prepositioned disaster-response facilities (DRFs) in terms of population in order to minimize the distance between demand points and DRFs considering the earthquake risk. The DRFs of the new container warehouses proposed by AFAD (Turkish Prime Ministry Disaster and Emergency Management Presidency), Turkish Red Crescent warehouses, and AFAD Civil Defense Search and Rescue City Directorates are considered in this study. Turkish Red Crescent Society is a humanitarian organization that provides relief to the vulnerable and those in need by mobilizing the power and resources of the community, and AFAD is the government agency concerning disasters and emergencies, and works like an umbrella organization, collaborating with the Ministry of Foreign Affairs, the Ministry of Health, the Ministry of Forests and Hydraulic Works, and other relevant ministries as well as nongovernmental organizations. We develop a mathematical model that determines the assignment of each demand point to each DRF by restricting the destruction powers and restricting the capacities of each DRF with its population size.

The rest of the chapter is organized as follows: In Section 7.2, we provide an overview of prepositioning in humanitarian logistics and risk management in disasters. In Section 7.3, we describe the system and problem in detail and present an integer programming model formulation. In Section 7.4, we test the model with case studies and report the computational results. Finally, we conclude and discuss future work in Section 7.5.

7.2 Literature Review

Despite humanitarian logistics' importance, the literature in this area is limited (Van Wassenhove 2006). Altay and Green (2006) survey the literature to identify potential research directions in disaster operations, discuss relevant issues, and provide a starting point for interested researchers.

In the fall of 2005, since hurricanes Katrina, Wilma, and Rita caused damage of more than $100 billion and highlighted the inadequacy of existing preparedness strategies, some research effort was aimed at devising prepositioning plans for emergency supplies (Rawls and Turnquist 2010).

Ukkusuri and Yushimoto (2008) modeled the prepositioning of supplies as a location-routing problem. Their model incorporates the reliability of the ground transportation network in case of any destruction happened. They maximize the probability that all the demand points can be served by a service location given fixed probabilities of link/node failure and a specified budget constraint. This model is related to our study in terms of demand points and service locations.

Balcik and Beamon (2008) developed a model to design a prepositioning system that balances the costs against the risks in the relief chain, which is a variant of the maximal covering location model, integrates facility location and inventory decisions, considers multiple item types, and captures budgetary constraints and capacity restrictions. It is revealed by the results of computational experiments that there are effects of pre- and postdisaster relief funding on relief system's performance, specifically on response time and the proportion of demand satisfied.

Duran et al. (2011) developed a mixed-integer programming inventory-location model to find the optimal configuration while considering a set of typical demand instances given a specified upfront investment (in terms of the maximum number of warehouses to open and the total inventory available to allocate) to determine the configuration of the supply network that minimizes the average response time over all the demand instances all over the world. The model obtains the typical demand instances from historical data; the supply network consists of the number and the location of warehouses and the quantity and type of items held in inventory in each warehouse. The basic differences between this study and our study are stock prepositioning, response times, and coverage area since our model provides an emergency response by assigning demand points to the DRFs with minimum earthquake risk in Turkey.

Görmez et al. (2011) developed a mathematical model to determine the locations of DRFs for Istanbul with the objectives of minimizing the average-weighted distance between casualty locations and DRFs, and opening a small number of facilities, subject to distance limits and backup requirements under regional vulnerability considerations. They analyzed the trade-offs between these two objectives under various disaster scenarios and investigated the solutions for several modeling extensions. The main difference of our study is our aim of covering all of Turkey and considering a single objective of minimizing total traveled distance.

Dükkancı et al. (2011) developed a model for Turkish Red Crescent Society (i.e., Kızılay in Turkish) that determined the DRF locations by evaluating demographic and past disasters' information to cover maximum number of people.

Risk is a widely used term in everyday life and businesses. Knight (1921) defined risk as "if you don't know the for sure what will happen, but you know the odds, that's risk, and if you don't even know the odds, that's uncertainty." The concept of resilience is closely related to the capability and ability of an element to return to a predisturbance state after a disruption (Bhamra et al. 2011). After the disaster, there might be risks related to the disruption of transportation roads and long delivery time, which should be well analyzed. In this study, we used an earthquake risk map, including destruction powers to integrate risk concept into our model.

To the best of our knowledge, the assignment of demand points to prepositioned DRF locations (in terms of cities) throughout Turkey considering that the earthquake risk has not been analyzed thoroughly. The next section presents an integer programming model for assigning city demand points to prepositioned DRF locations in Turkey considering the earthquake risk.

7.3 Solution Methodology

When the prepositioning literature is analyzed, it is seen that either the distance traveled between DRFs and affected areas or elapsed time is minimized by considering the closeness of DRFs to the disaster-prone areas. In this study, the affected areas by the disaster are called as *demand points*. The assumptions used in the problem are given in the following:

- The DRFs can cover a maximum 15,000,000 population, because we limit the coverage with the population sizes of the cities that have DRFs.
- The DRFs can satisfy their own requirements from an infinite supply.

7.3.1 Mathematical Model

The objective is to minimize the distances between demand points and DRFs in order to quickly respond to the requirements of beneficiaries. The following notation is used for the DRF assignment model:

Sets

C set of DRF locations; $i \in C$
T set of demand points; $j \in T$

Parameters

D_{ij}: Distance between DRF i and demand point j
K_j: Population of demand point j
P_i: Capacity of DRF i in terms of population
R_{ij}: Average destruction power based on the magnitude of the earthquake for DRF i and demand point j

Decision Variables

$$x_{ij} = \begin{Bmatrix} 1 & \text{if demand point } j \text{ is covered by DRF } i \\ 0 & \text{Otherwise} \end{Bmatrix}$$

The mathematical model for the problem is as follows:

$$\min \sum \sum D_{ij} x_{ij} \tag{7.1}$$

subject to

$$\sum K_j x_{ij} \leq P_i \quad \forall i \in C \tag{7.2}$$

$$\sum x_{ij} \geq 1 \quad \forall j \in T \tag{7.3}$$

$$\sum_j R_{ij} x_{ij} \geq 1 \quad \forall i \in C \tag{7.4}$$

$$x_{ij} \in \{0,1\} \quad \forall i \in C, \quad \forall j \in T \tag{7.5}$$

The objective function (7.1) minimizes the distance between the DRFs and demand points. Constraint set (7.2) ensures that a DRF can cover the population of a demand point j up to its population capacity. Constraint set (7.3) ensures that every demand point must be covered by at least one facility. Constraint set (7.4) satisfies that the total average destruction power between DRFs and demand points must be greater than or equal to one. Thus, the DRFs cover the demand points that have large destruction powers. Constraint set (7.5) ensures that the location coverage variables are binary.

7.4 Experimental Studies

The proposed mathematical model is tested for DRFs of the new container warehouses proposed by AFAD, Turkish Red Crescent warehouses, and AFAD Civil Defense Search and Rescue City Directorates in the following sections. Only the data set and computational results of the first case will be given in detail, and the visual representation of the results will be given for others. The data set (i.e., risk, population, distance) used for all these cases are the same.

7.4.1 First Case

This experiment is conducted for 27 container warehouse locations proposed by AFAD recently. Earthquake risk data are taken from the earthquake risk map at city and town level, which was prepared by Prof. Dr. Ahmet ERCAN (Ercan 2010). The distances between cities are taken from KGM (General Directorates for Highways 2013). Demographic information of cities and towns (populations) is taken from TUIK (Turkish Statistical Institute 2012).

The average destruction powers given in Table 7.1 are derived from the minimum and maximum destruction powers in the earthquake map (Ercan 2010). The first column of the table shows cities, the second column shows the populations, and the third column shows the corresponding risk regions. Fourth and fifth columns show minimum and maximum destruction powers corresponding to risk regions. The sixth column is the average destruction power value calculated by taking average of minimum and maximum destruction powers. This is taken as the average to have a moderate representation of the

Table 7.1 Sample from the Data Set

CITY	2012 POPULATION	RISK REGION	MIN. DESTRUCTION POWER (A-CM/SN²)	MAX. DESTRUCTION POWER (A-CM/SN²)	AVG. DESTRUCTION POWER (A-CM/SN²)
Adana	2,125,635	IX	0.31	0.71	0.51
Adıyaman	595,261	IX	0.31	0.71	0.51
Afyon	703,948	IX	0.31	0.71	0.51
Ağrı	552,404	XI	1.50	3.1	2.30
Amasya	322,283	X	0.71	1.50	1.10
Ankara	4,965,542	VIII	0.15	0.31	0.23
Antalya	2,092,537	IX	0.31	0.71	0.51
Artvin	167,082	VIII	0.15	0.31	0.23
Aydın	1,006,541	X	0.71	1.50	1.10
Balıkesir	1,160,731	X	0.71	1.50	1.10
...					
Kilis	124,320	VI	0.03	0.07	0.05
Osmaniye	492,135	VII	0.07	0.15	0.11
Düzce	346,493	XII	3.10	7.10	5.10

destruction power. According to Table 7.1, the maximum average destruction power is 7.1g for Düzce in the most risky area (XII). The minimum average destruction power is 0.051g for Kilis in the least risky area (VI).

The proposed mathematical model was solved using GAMS 23.7 with CPLEX 11 Solver. The total traveled distance is 10,778 km with 59 (i, j) pairs. The (i, j) pair stands for the assignment of demand point j to DRF i. We identified them as pair since our model determines the (i, j) pair, and we make comparison among each cases by the pair assignments. The total average destruction power between (i, j) pairs is 60.92.

The assignment of demand points to DRFs is given in Table 7.2 for this case. In the first and fifth columns, the prepositioned DRFs are listed. In the second and sixth columns, the assigned demand points to DRFs are listed. In the third and seventh columns, the distances between the DRFs and the demand points are given as (i, j) pairs. In the fourth and eighth columns, the average destruction power between (i, j) pairs is given. The results show that demand points are assigned to DRFs with an ability to serve the demand points in at most 4 h by highways in normal conditions except for Elazığ-Rize assignment with 570 km. It can be concluded that each DRF covers at

Table 7.2 Assignment of Demand Points to DRFs for Container Warehouses Proposed by AFAD

DRFS	COVERED DEMAND POINTS	DISTANCE BETWEEN (*I*, *J*) PAIRS	AVG. DESTRUCTION POWER BETWEEN (*I*, *J*) PAIRS	DRFS	COVERED DEMAND POINTS	DISTANCE BETWEEN (*I*, *J*) PAIRS	AVG. DESTRUCTION POWER BETWEEN (*I*, *J*) PAIRS
Adana	Mersin	69	0.37	Manisa	Aydın	156	0.81
	Niğde	205	0.31		Uşak	195	0.51
	Karaman	289	0.37	Kahramanmaraş	Gaziantep	80	0.17
Adıyaman	Bingöl	349	0.81		Tokat	415	0.67
	Şanlıurfa	110	0.31		Osmaniye	100	0.17
Afyon	Eskişehir	144	0.37	Muğla	Aydın	99	0.81
	Kütahya	100	1.41		Isparta	292	0.51
Ankara	İstanbul	453	1.27	Muş	Bitlis	83	0.51
Antalya	Burdur	122	0.51		Siirt	180	0.31
	Isparta	130	0.51		Şırnak	275	0.51
Balıkesir	Kütahya	224	1.70	Samsun	Giresun	196	1.27
Bursa	İzmir	322	2.00		Ordu	152	1.27
Denizli	Aydın	126	0.81		Sinop	163	1.27
	Uşak	150	0.51	Sivas	Amasya	222	1.11
Diyarbakır	Bingöl	144	0.81		Kayseri	195	0.67
	Mardin	95	0.31	Tekirdağ	Çanakkale	188	1.41
	Batman	100	0.31		Edirne	140	1.27
Elazığ	Malatya	98	0.37		Kırklareli	121	1.27

(*Continued*)

Table 7.2 (*Continued*) Assignment of Demand Points to DRFs for Container Warehouses Proposed by AFAD

DRFS	COVERED DEMAND POINTS	DISTANCE BETWEEN (*I*, *J*) PAIRS	AVG. DESTRUCTION POWER BETWEEN (*I*, *J*) PAIRS	DRFS	COVERED DEMAND POINTS	DISTANCE BETWEEN (*I*, *J*) PAIRS	AVG. DESTRUCTION POWER BETWEEN (*I*, *J*) PAIRS
	Rize	570	2.67	Van	Hakkari	202	1.41
Erzincan	Gümüşhane	131	2.61		Iğdır	225	1.27
	Trabzon	231	2.67	Aksaray	Çorum	326	0.61
	Tunceli	130	2.81		Kırşehir	110	0.17
Erzurum	Ağrı	184	1.41		Konya	148	0.17
	Artvin	226	0.37		Nevşehir	75	0.11
	Kars	203	0.51	Kırıkkale	Çankırı	105	0.37
	Bayburt	125	0.31		Çorum	167	0.67
	Ardahan	230	0.37		Yozgat	141	0.23
Hatay	Kilis	147	1.18	Yalova	Bilecik	129	1.21
Kastamonu	Bartın	181	0.67	Düzce	Bolu	45	3.10
	Karabük	114	0.67		Zonguldak	114	2.67
Kocaeli	Sakarya	37	5.10				

least one demand point and at most five demand points such as Bursa and Erzurum DRFs. The demand points receive relief supplies from one facility since the facility sizes are limited with their population sizes. Few demand points receive relief supplies from more than one facility like Kütahya, Aydın, Uşak, and Bingöl.

The demonstration of the assignments for (i, j) pairs is given in Figure 7.1. It shows the assignment of demand points to DRFs, which are symbolized by a container.

7.4.2 Second Case

This experiment is conducted for 30 Turkish Red Crescent warehouses. The proposed mathematical model was solved using GAMS 23.7 with CPLEX 11 Solver. The total distance traveled is 10,617 km with 59 (i, j) pairs. The total average destruction power between (i, j) pairs is 47, which is less than the value observed in the first case. The visual representation of the assignment of demand points to DRFs is given in Figure 7.2 for the second case. As seen from Figure 7.2, the demand points are assigned to DRFs with an ability to serve the demand points in at most 4 h by highways in normal conditions except for Gaziantep-Çorum and Rize-Amasya assignments with 630 and 535 km, respectively. Each DRF covers at least one demand point and at most five demand points such as Ağrı and Gaziantep DRFs. The demand points receive relief supplies from one facility since the facility sizes are limited with their population sizes. Few demand points receive relief supplies from more than one facility like Kütahya, Çankırı, Aydın, Bitlis, and Bingöl.

7.4.3 Third Case

This experiment is conducted for 11 DRFs of AFAD Civil Defense Search and Rescue City Directorates. The total traveled distance is 13,997 km with 71 (i, j) pairs. The total distance traveled is higher than the first and second case studies because the number of DRFs is fewer. The total average destruction power between (i, j) pairs is 72.71, which is more than the observed value in the first and second case studies. The visual representation of the assignment of demand points to DRFs is given in Figure 7.3 for this case. As seen from

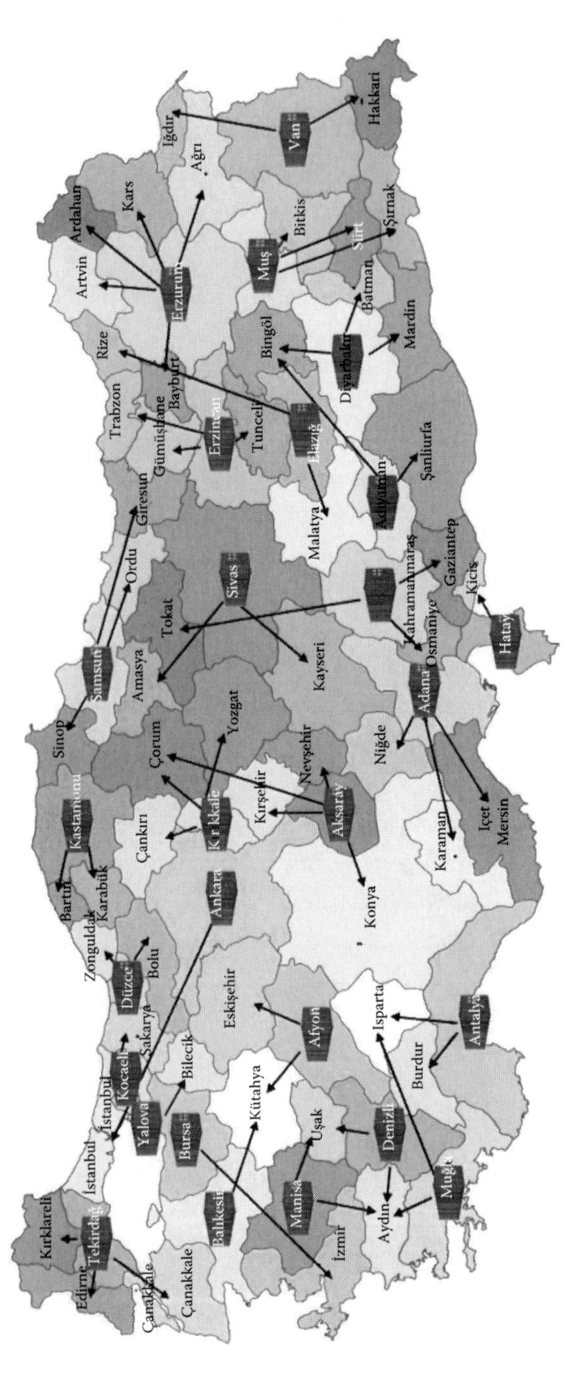

Figure 7.1 Assignment of cities for container warehouses proposed by AFAD.

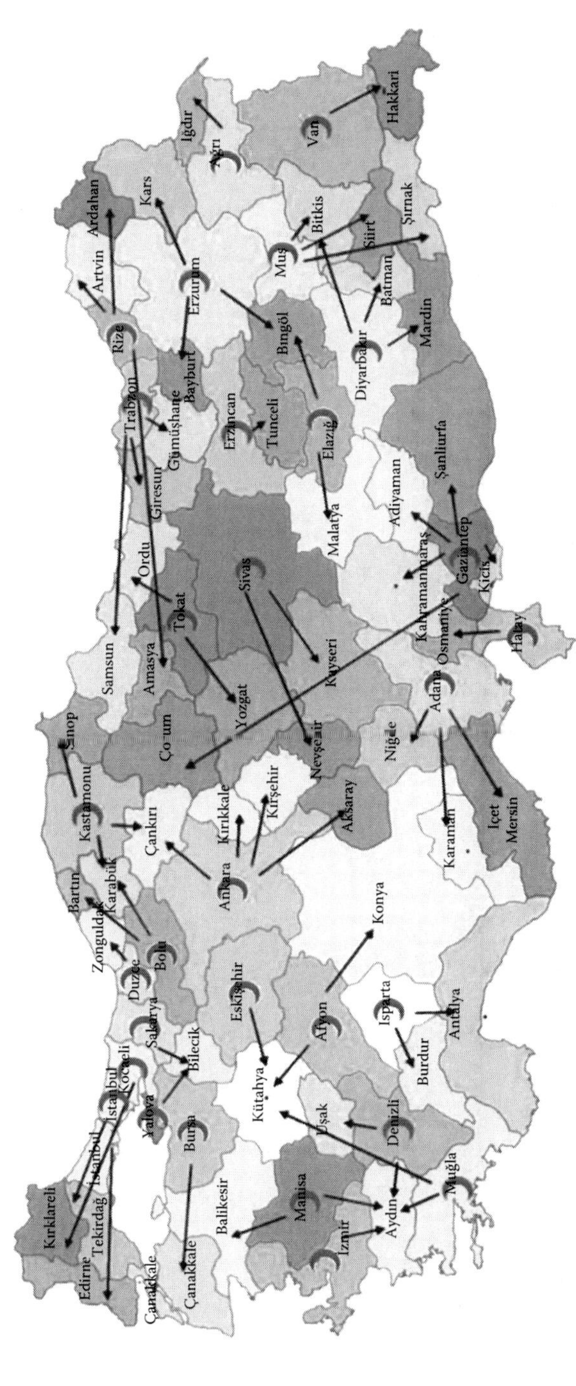

Figure 7.2 Assignment of cities for Turkish Red Crescent warehouses.

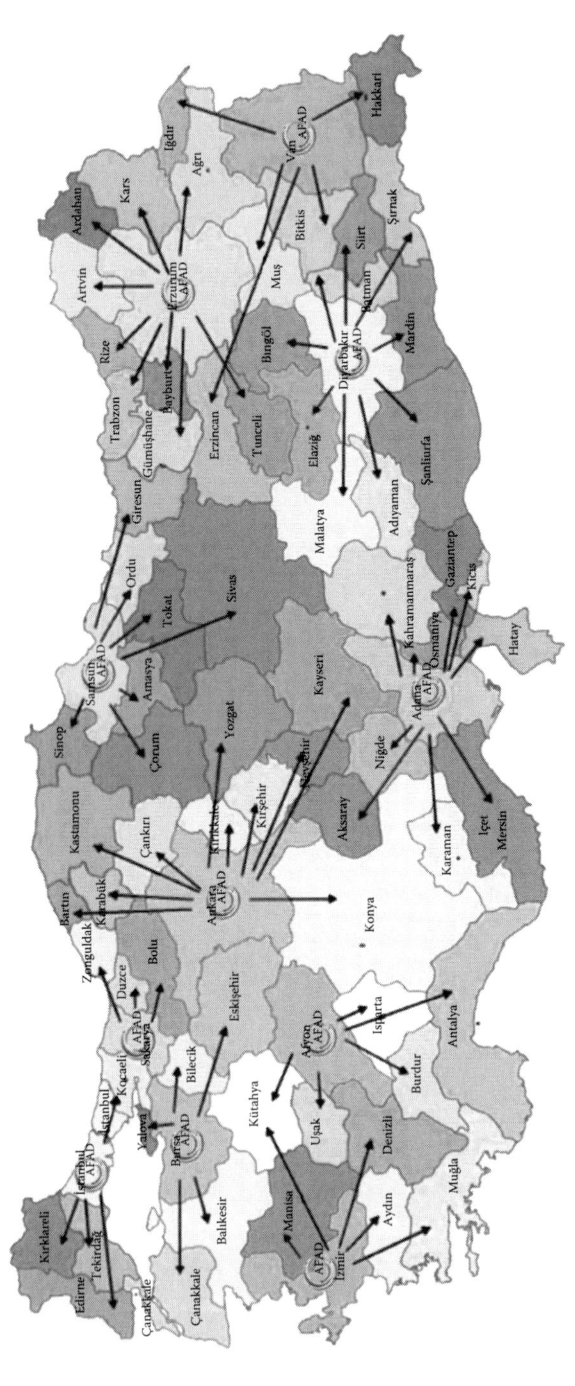

Figure 7.3 Assignment of cities for AFAD Civil Defence and Rescue City Directorates.

Figure 7.3, the demand points are assigned to DRFs with an ability to serve the demand points in at most 4 h by highways in normal conditions except for Van-Erzincan assignment with 602 km. Each DRF covers at least 1 demand point and at most 12 demand points such as Diyarbakır DRF. The demand points receive relief supplies from one facility since the facility sizes are limited with their population sizes. Kütahya is an exception since it receives relief supplies from two facilities.

The set of the warehouses used in the first, second, and third cases are given in Figure 7.4 by displaying the overlapping DRFs among them. There are 12 overlapping cities for container warehouses and Turkish Red Crescent warehouses, 3 overlapping cities for AFAD warehouses and Turkish Red Crescent warehouses, and 1 overlapping city for container warehouses and AFAD warehouses. Eight cities belong to only Turkish Red Crescent warehouses, and seven cities

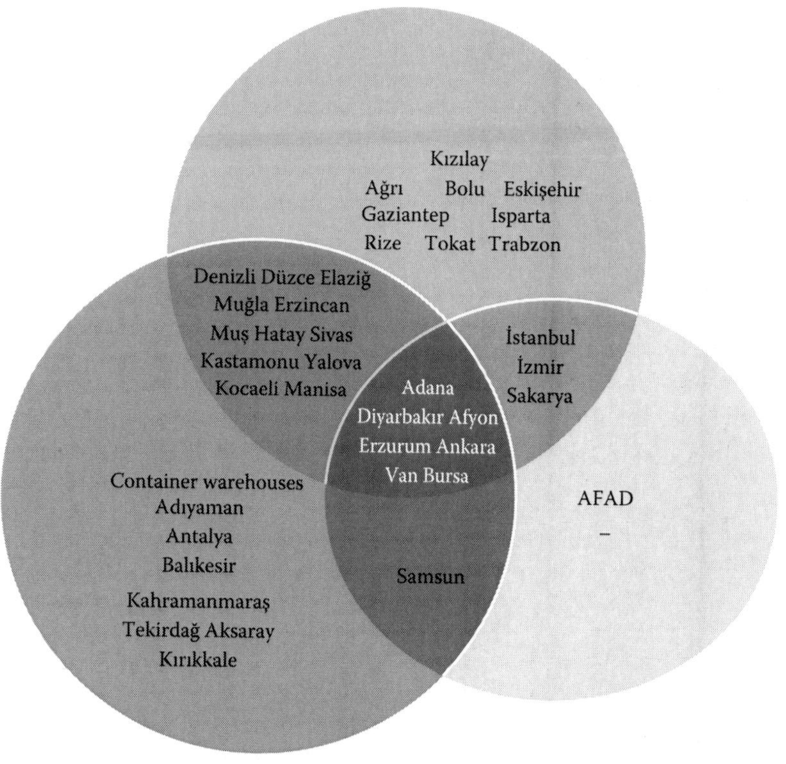

Figure 7.4 The set of the warehouses used in the case studies.

Table 7.3 Comparison of the Cases according to Numerical Results

	(A)	(B)	(C)	(D)	(E)	(F) = (E)/(B)
CASE	NO. OF DRFS	NO. OF (*I*, *J*) PAIR	NO. OF DEMAND POINTS COVERED BY TWO DRFS	TOTAL TRAVELED DISTANCE (KM)	TOTAL AVG. DESTRUCTION POWER	AVG. DESTRUCTION POWER OF (*I*, *J*) PAIR
First case	27	59	5	10,778	60.92	1.033
Second case	30	59	8	10,617	47.00	0.797
Third case	11	71	1	13,997	72.71	1.024

belong to only container warehouses. Seven DRFs are common in all cases: Adana, Diyarbakır, Afyon, Erzurum, Ankara, Van, and Bursa.

The summary of three cases is depicted in Table 7.3 for comparison. In the first column of the table, three cases are given for container warehouses, Turkish Red Crescent warehouses, and AFAD warehouses, respectively. In the second and third columns, the number of DRFs belonging to each case and the number of (i, j) pair by the assignment model are shown. In the fourth column, the number of demand points covered by more than one DRF is given. In the results of the model for three cases, we observed that five, eight, and one demand points are covered by two DRFs. The coverage by two DRFs is induced by the model parameters and could be increased when the capacity limits of the DRFs are increased. The fifth and sixth columns are for the total distance traveled and total average destruction powers obtained by the assignment model. In the last column of Table 7.3, the average destruction power of (i, j) pair for each cases is calculated by dividing the total average destruction power to the number of (i, j) pair obtained in the result of the assignment model. Thus, the average destruction power is found per (i, j) pair. This value could be compared with the situation when it is thought as there are 81 DRFs (i.e., one warehouse in each city) and 81 demand points. If each demand point is assigned to each DRF, then we have 81×81 assignment, and the overall average destruction power per assignment is found as 0.85 by dividing the average destruction powers of each (i, j) pair to the number of demand points, which is 81. This means that when all cities behave like DRFs and are able to serve to all cities, the average destruction value of any assignment is 0.85. However, we take into account the population capacity of each DRF as well as destruction powers. It can be said that the assignment of demand points to the prepositioned DRFs are less risky when the obtained value is less than 0.85, so 0.85 is taken as a *moderate value*. When considered from this point of view, the second case is superior to the other cases, and it has the least average destruction power per (i, j) pair.

7.5 Conclusion and Future Work

In this study, our aim was to minimize the total distance between prepositioned DRFs and the demand points in cities by considering facility capacities and the average earthquake destruction powers between them. We developed an integer programming model for the assignment of demand points to the prepositioned DRFs. We tested our model with three cases, namely, container warehouses proposed by AFAD (Turkish Prime Ministry Disaster and Emergency Management Presidency), Turkish Red Crescent warehouses, and AFAD Civil Defense Search and Rescue City Directorates. In the results, we obtained the total distance traveled, the number of covered demand points by each DRF, and the total average earthquake destruction power. In the study, we observed that humanitarian relief organization considered in experimental studies has common cities to store the relief items being unaware of the warehouse decisions of each other. It shows that those common cities are suitable to have DRFs. This also reveals that some of the factors they consider in selecting the DRF locations are the same.

This study can be utilized to see the assignment effects on the average destruction powers and the number of the assigned demand points. In the study, we observed the total average destruction powers for each case, and we observed that they have different average destruction powers per (i, j) pair, two of them are above the moderate value, which is 0.85, and one of them is below the moderate value. The assignment for Turkish Red Crescent warehouses is the best in terms of all performance measures.

The study can be extended by considering the exact locations and capacities of DRFs. In future studies, the distances of exact locations for DRFs would support the implementation of the model and improve the analysis. Backup facility concept can be introduced to the model in order to be safe in the risks of not delivering the relief items to demand points when the warehouse or roads are destroyed.

Acknowledgments

We are grateful to Turkish Prime Ministry Disaster and Emergency Management Presidency (AFAD) personnel and the Scientific and Technological Research Council of Turkey (TUBITAK) who partially supported this study with the research Grant No. 113M493.

References

AFAD. T.C. başbakanlık Afet ve Acil Durum Yönetimi Başkanlığı [cited December 2012]. Available from http://www.afad.gov.tr.

Altay, N. and W. G. Green. 2006. OR/MS research in disaster operations management. *European Journal of Operational Research* 175(1): 475–493.

Balcik, B. and B. M. Beamon. 2008. Facility location in humanitarian relief. *International Journal of Logistics: Research and Applications* 11(2): 101–121.

Bhamra, R., S. Dani, and K. Burnard. 2011. Resilience: The concept, a literature review and future directions. *International Journal of Production Research* 49(18): 5735–5793.

Dükkancı, O., Ö. Koşak, A. İ. Mahmutoğulları, H. Özlü, and N. Timurlenk. 2011. Merkezi Afet Yönetiminde Karar Destek Sistemi Tasarımı. Bilkent University, Ankara, Turkey.

Duran, S., M. A. Gutierrez, and P. Keskinocak. 2011. Pre-positioning of emergency items worldwide for CARE international. *Interfaces* 41(3): 223–237.

Ercan, A. Ö. S. 2010/12. Türkiye'nin Deprem Çekincesi: İl il, ilçe ilçe Deprem Belgeseli. PARA Dergisi, 14–20 Mart.

Görmez, N., M. Köksalan, and F. S. Salman. 2011. Locating disaster response facilities in Istanbul. *Journal of the Operational Research Society* 62(7): 1239–1252.

KGM. General Directorates for Highways [cited December 2013]. Available from http://www.kgm.gov.tr.

Knight, F. H. 1921. *Risk, Uncertainty and Profit.* Hart, Schaffner & Marx, Boston, MA.

Rawls, C. G. and M. A. Turnquist. 2010. Pre-positioning of emergency supplies for disaster response. *Transportation research part B: Methodological* 44(4): 521–534.

Thomas, A. S. and L. R. Kopczak. 2005. From logistics to supply chain management: The path forward in the humanitarian sector. Fritz Institute, San Francisco, CA.

TUIK. Turkish Statistical Institute [cited December 2012]. Available from http://tuik.gov.tr.

Ukkusuri, S. V. and W. F. Yushimoto. 2008. Location routing approach for the humanitarian prepositioning problem. *Transportation Research Record: Journal of the Transportation Research Board* 2089(1): 18–25.

Van Wassenhove, L. N. 2006. Humanitarian aid logistics: Supply chain management in high gear. *Journal of the Operational Research Society* 57(5): 475–489.

8

Factors Affecting the Purchasing Behaviors of Private Shopping Club Users

A Study in Turkey

SERCAN AKKAŞ, CEREN SALKIN, AND BAŞAR ÖZTAYŞI

Contents

8.1 Introduction

Since technological improvements have provided advantages of convenience in business and as the Internet's widespread usage is increasing all around the world, existing companies have started adapting and integrating their existing business models to the electronic environment with recent entrepreneurs; the result, obviously, is that novel business models have started to appear. One of these is *online shopping*, which is growing in Turkey as well as all over the world. As a result, modern business models that have come into existence have brought in new perspectives in the e-commerce sector due to improvements in online shopping. In recent years, "Private

Shopping Clubs (PSCs)" have emerged as a new type of business model with branch categorizations on online shopping websites, making them popular business models that encourage fast shopping for customers.

PSCs have their own internal business dynamics and characteristics as is found in every business model. PSCs are websites on which products are classified in accordance with certain sectors and brands are sold online in smaller amounts with huge discounts rather than mass customization. These discounts have time limitations to induce rapid shopping and increase of sales. Besides, members are notified before entering their discounts. This online business model works usually by invitation and is structured as a closed-loop shopping system. People who do not have membership in related PSCs will not be able to see the products; nor will they be informed about the daily discounts and promotions.

Previous studies that examined the purchasing behavior of customers in online shopping proposed different models for determining end users' shopping behavior. In this study, the PSC model is examined with the Technology Acceptance Model (TAM). The recommended model includes factors such as the tendency for utilitarian and hedonic shopping behaviors, social effects, personal innovativeness, e-commerce service quality, among others. The study focuses on explaining the behaviors of PSC users in order to make improvements with respect to an increase in online shopping. Data are analyzed with multiple regression analysis and the proposed model is confirmed within an acceptable error rate. Results of the study can be utilized as a guide for existing PSCs and recent enterprises with the intention of engaging this new business model for understanding the purchasing tendencies of the customers. Researchers can also examine what PSC users expect from online shopping websites and what kind of properties they give importance to in the shopping process. Moreover, the study could be useful in terms of investigating the intention behind the purchasing behavior. In addition to this, it could also analyze the customer satisfaction at the end of their actual purchase in order to improve on innovative models in related topics.

This chapter is organized as follows. Section 8.2 provides information about online shopping, e-commerce success factors and previous studies in the literature. Section 8.3 consists of the statistical formulization of the proposed model. The section also presents the results of the conducted survey. In Section 8.4, suggestions for further research is given and finally the paper is concluded in Section 8.5.

8.2 Background

In this section, we investigate the critical factors of purchasing from online shopping websites and e-commerce applications. First of all, some information on e-commerce and online shopping perspective is given. Then, previous studies are mentioned for understanding the concept.

8.2.1 E-Commerce and Online Shopping

Online shopping is a continuous and integrated process that begins with attaining the materials and ends with the final customer's purchase of goods or the final customer's after-sales service (Nath et al., 1998). First of all, the behavior of online shopping players alters continuously in accordance with other players. For instance, seller's main objective is to sell raw material or semi-finished products in order to make a better profit while the customer's aim is to buy goods at a cheap rate. In addition to these diverse purposes, online shopping is affected by the realities of world events that could cause uncertainties (Navarro et al., 2007). In this dynamic environment, B2B and B2C members should think about other chain members' decisions and make contacts regarding the coordination mechanism with respect to external economic changes and environmental factors that could affect the whole chain (Numberger and Rennhak, 2005). Therefore, Peppers and Rogers (2001) referred a centralized purchasing management model for accurate decision making. Besides that, purchasing risk management and defining risk performance criteria became key elements for purchasing activities (Lee et al., 2008). In this respect, Lee et al. (2003) proposed a purchasing management concept consisting of five components: triggers, decision-making characteristics, management factors, management responses, and performance outcomes. They emphasized that aggregate management comprising nodes and chains diversified activities that were combined with e-commerce management activities. According to Hsia et al. (2008), there is a necessity for activities that reduce the flow in purchasing management to be controlled and corrected simultaneously. Hawes and Lumpkin (1986) described uncertain parameters in purchasing management such as demand, supply, processing, transportation, lack of goods, and capacity. Choudhury and Hartzel (1998) noted performance control parameters in e-commerce applications from the perspective of resources, lead times, capacity, and inventory levels. According to Choudhury and Hartzel (1998), e-commerce consists of five factors: delivery lateness, price could be above expectations, quality failures, confidence, and flexibility problems in the literature. Besides all these factors related with e-commerce activities, there is lack of physiological perspective in relation to purchasing and the tendency for Internet shopping, as mentioned by the authors.

8.2.2 E-Commerce Success Factors and Previous Modeling Approaches

In the literature, models used for various purposes generally utilize different optimization tools such as linear programming, stochastic modeling, and deterministic modeling (Atchariyachanvanich et al., 2007). Because of the shortcomings in the interpretation of the results, they do not provide sufficient information about purchasing behavior and the measure of e-commerce success (Aboelmaged, 2010). Additionally, some authors also mention that there is a lack of models that can analyze interactions among the physiological components, which should be viewed (Anumba and Ruikar, 2002). Due to the emerging and dynamic virtual relations between these factors, models should acquire sudden changes and provide continuous monitoring. From the point of view of this dynamic, mathematical models that formulize purchasing behavior were unable to cope with the difficulties of instant variations and continuous monitoring of the entire system (Browne and Cudeck, 1993). In addition to this static structure, interactions and variations of the parameters could not be shown in the mathematical models. Therefore, the models constructed by system dynamics, agent-based simulation, and case-based reasoning are more suitable for determining causal relationships and controlling parameter effects on system behavior (Jiang and Qian, 2009).

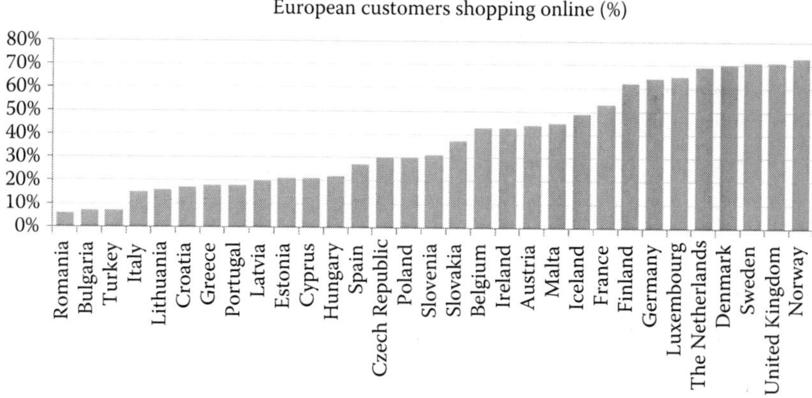

Figure 8.1 Online shopping rates in Europe in 2012. (Eurostat Report—Online Payments 2012; http://epp.eurostat.ec.europa.eu/cache/ITY_OFFPUB/KS-CD-12-001/EN/KS-CD-12-001-EN.PDF.)

Table 8.1 E-Commerce Success

CONTRIBUTION TO SUCCESS	EFFECTS	
	INTERNAL EFFECTS	EXTERNAL EFFECTS
Triggers	1. Cost management 2. Reputation 3. Market 4. Entrance to the sector	1. Product pricing 2. Devoted time 3. Convenience 4. External relationships
Obstacles	1. Financial predicaments 2. Risks 3. Experience	1. Costs caused from customers 2. Delivery time 3. Transaction risk 4. Access

Source: Quaddus, M. and Achjari, D., *Telecommun. Policy*, 29, 127, 2005.

In recent years, online shopping has progressed as an effective management approach that regards unstable external factors and interactions among them (Figure 8.1). The main question is how key performances related with e-commerce applications (Table 8.1) interact with each other and how these interactions affect the whole online shopping processes (Jun et al., 2004).

In relation to this problem, purchasing behavior originates from three components: information on a loss-making event, probability of existing risks, and effect of the event. Purchasing management is a systematic way of management that includes determining uncertainties in the processes that could cause failures and trouble in a system and monitoring these processes and indicators continuously

(Ho et al., 2007). In this sense, e-commerce management should incorporate financial risks, which imply a balance of cost and profit; demand uncertainty and fluctuations, which cause the butterfly effect; and tardiness and delays (Chen et al., 2008).

All these factors trigger purchasing behavior and affect sales in online shopping. Since online shopping websites deal with a huge number of complaints, this study investigates psychological factors that indirectly affect purchasing behavior. Unlike former studies, this model involves individual perspectives of the purchasing behavior as the main indicators in the analysis.

8.3 Technology Acceptance Model with Multiple Regression Analysis

Technology Acceptance Model (TAM) was first introduced by Davis in 1989 in his doctorate thesis. This methodology is based on the Theory of Reasoned Action. This model is evaluated as a sociophysiological method that has a well-structured perspective of forecasting the acceptability of new technologies (Wang, 2002). The basis of this theory involves a cost–benefit paradigm, an adaptation of innovations, self-sufficiency, and a theory of expectations theory (Davis, 1989). Additionally, this model has the capability of making improvements in the model with respect to other factors that could affect the tendency of purchasing (Legris et al., 2003).

Substantial factors that impress upon purchasing intuition are listed as follows:

- External variations such as late delivery, high prices
- Perceived usefulness
- Perceived ease of use
- Attitude to utilization
- Intuitions of use
- Usage of the system

The basic infrastructure of the model includes the following factors:

- Utilitarian and hedonic shopping orientation
- Information quality related to the product
- Reliance on the website
- Website quality

- Personal innovativeness
- Perceived pleasure
- E-commerce service quality
- Content wealth
- Satisfaction

For measuring the direct and indirect effects on the purchasing behavior of the customers in private shopping, we construct a multiple regression model that can be shown generally as follows:

$$Y = \beta_0 + \beta_1 x_1 + \beta_2 x_2 + \beta_3 x_3 + \beta_4 x_4 + \cdots$$

In this equation, Y denotes the dependent variable that symbolizes private shopping behavior and $x_1, x_2, \ldots x_n$ are independent variables that directly or inversely affect Y. The independent variables are determined as perceived usefulness, perceived ease of use, attitude to utilization, intuitions of use, usage of the system, utilitarian and hedonic shopping orientation, information quality related to the product, reliance on the website, website quality, personal innovativeness, perceived pleasure, e-commerce service quality, content wealth, and satisfaction. In order to demonstrate the effect of these variables, a questionnaire is applied in accordance with the Likert scale. The demographic characteristics of the survey are given in Tables 8.2 through 8.4.

Before using multiple regression analysis, reliability analysis should be applied to assess the suitability of the data to the statistical analysis. We can examine the reliability of the data by checking the Cronbach Alfa value. If the Cronbach Alfa value is greater than or equal to 0.7, one may conclude that the data are suitable for statistical analysis. The Cronbach Alfa values are represented in Table 8.5. In this table, the utilitarian and hedonic shopping orientation and individual norms are

Table 8.2 Considered Branches in the Survey

PRIVATE SHOPPING CLUBS	FREQUENCY	PERCENTAGE (%)
Markafoni	304	41.87
Trendyol	130	17.91
Limango	44	6.06
Morhipo	30	4.13
1v1y	20	2.75
Others	198	27.27

Table 8.3 Demographic Properties of the Participants

CRITERIA	CHARACTERISTICS	FREQUENCY	PERCENTAGE (%)
Sex	Male	342	47.11
	Female	384	52.89
Marital status	Married	78	10.74
	Single	648	89.26
Age	<25	416	57.30
	25–40	286	39.39
	>40	24	3.31
Education	Doctorate	7	0.96
	MSc	105	14.46
	BSc	558	76.86
	Undergraduate	7	0.96
	Technical High School	8	1.10
	High School	37	5.10
	Secondary School	4	0.55
Monthly income	<1000 TL	165	22.73
	1000 TL–1500 TL	135	18.60
	1501 TL–2500 TL	196	27.00
	2501 TL–5000 TL	204	28.10
	>5000 TL	26	3.58

Table 8.4 Previous Online Shopping Experience of the Participants

CRITERIA		FREQUENCY	PERCENTAGE (%)
Previous online shopping experience	Shopping less than once in 1–2 years	81	11.16
	Shopping once or twice a year	91	12.53
	Shopping three or four times in a year	196	27.00
	Shopping once or twice in a month	253	34.85
	Shopping two or three times in a month	105	14.46

not reliable for multiple regression analysis. If these items are deleted, the reliability of the analysis will increase.

After the reliability analysis, we design the multiple regression model as follows (Figure 8.2): H_0 denotes the main hypothesis which demonstrates that the changes made in the population cannot affect the arithmetic average of the universe. H_i indicates the counterpart situation.

Subsequent to this, we group the independent variables in order to construct the relationships between independent and dependent variables. As seen in Table 8.6, we gather 12 groups that can explain the variability of 69.275% with independent variables (Tables 8.7 and 8.8).

Table 8.5 Cronbach Alfa Values of the Variables

VARIABLE	CRONBACH ALFA VALUE IF ITEM IS DELETED	CRONBACH ALFA
Innovativeness	—	0.81
Utilitarian and hedonic shopping orientation	0.777	0.62
Individual norms	0.742	0.693
Perceived risk	—	0.813
Reliance on the website	—	0.924
Information quality related to the product	—	0.787
Website quality	—	0.78
E-commerce service quality	—	0.815
Content wealth	—	0.846
Attitude to utilization	—	0.877
Usage of the system	—	0.825
Perceived pleasure	—	0.903
Intuitions of use	—	0.847
Satisfaction	—	0.906

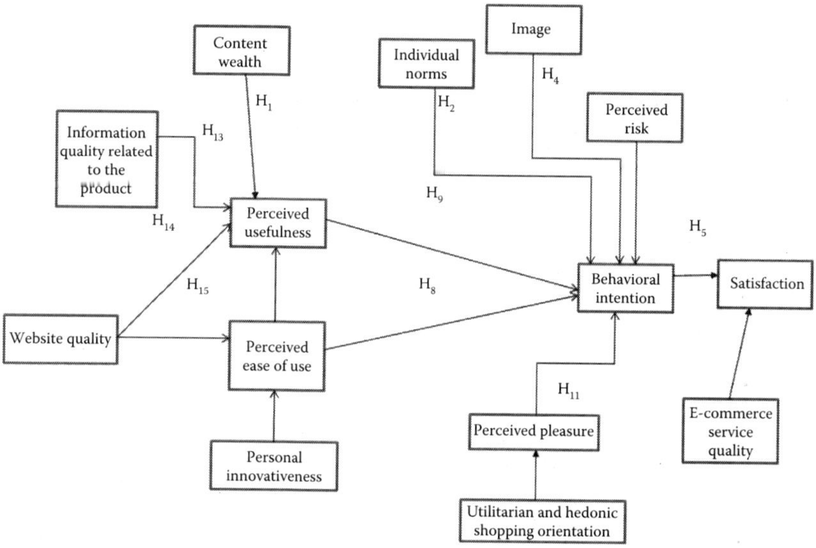

Figure 8.2 Multiple regression model for the explanation of purchasing behavior.

After applying the principal component analysis, we construct a multiple regression model in analysis of variance (ANOVA) by considering the significance value (p value) at 95% confidence level.

$$\text{Usage of the system} = 1.945 + (0.431 \times \text{website quality}) + (0.76 \times \text{personal innovativeness}).$$

Table 8.6 Principal Component Analysis Results

	INITIAL EIGEN VALUE			EXTRACTION SUMS OF SQUARES LOADINGS			ROTATION SUMS OF SQUARES LOADINGS		
COMPONENT	TOTAL	PERCENTAGE OF VARIANCE (%)	CUMULATIVE PERCENTAGE (%)	TOTAL	PERCENTAGE OF VARIANCE (%)	CUMULATIVE PERCENTAGE (%)	TOTAL	PERCENTAGE OF VARIANCE (%)	CUMULATIVE PERCENTAGE (%)
1	14,769	29.538	29.538	14.769	29.538	29.538	6.997	13.993	13.993
2	3,567	7.134	36.672	3.567	7.134	36.672	3.573	7.146	21.14
3	2,487	4.975	41.646	2.487	4.975	41.646	3.372	6.743	27.883
4	2,228	4.457	46.103	2.228	4.457	46.103	2.996	5.992	33.875
5	2,013	4.025	50.128	2.013	4.025	50.128	2.761	5.522	39.396
6	1,943	3.887	54.015	1.943	3.887	54.015	2.64	5.279	44.676
7	1,598	3.195	54.21	1.598	3.195	54.21	2.299	4.598	49.274
8	1,446	2.892	60.102	1.446	2.892	60.102	2.246	4.492	53.766
9	1,337	2.674	62.776	1.337	2.674	62.776	2.175	4.35	58.116
10	1,152	2.304	65.08	1.152	2.304	65.08	2.15	4.299	62.415
11	1,069	2.139	67.219	1.069	2.139	67.219	1.765	3.529	65.944
12	1,028	2.056	69.275	1.028	2.056	69.275	1.665	3.331	69.276

Table 8.7 Variables Indirectly Affecting Each Other

	NORM	RISK	SERVICE	PRAGMATIC	INNOVATIVENESS	WEBSITE	SATISFACTION	INTUITIONS OF USE	PERCEIVED PLEASURE	USAGE OF THE SYSTEM	ATTITUDE TO UTILIZATION
Satisfaction	0.12	−0.10	—	0.10	0.04	0.33	—	—	0.25	0.35	0.54
Intuitions of use	—	—	—	0.12	0.05	0.41	—	—	—	0.39	—
Perceived pleasure	—	—	—	—	—	—	—	—	—	—	—
Usage of the system	—	—	—	—	—	—	—	—	—	—	—
Attitude to utilization	—	—	—	—	0.07	0.35	—	—	—	—	—

Table 8.8 Variables Directly Affecting Each Other

	NORM	RISK	SERVICE	PRAGMATIC	INNOVATIVENESS	WEBSITE	SATISFACTION	INTUITIONS OF USE	PERCEIVED PLEASURE	USAGE OF THE SYSTEM	ATTITUDE TO UTILIZATION
Satisfaction		—	0.22	—	—	—	—	0.81	—	—	—
Intuitions of use	0.15	–0.13	—	—	—	—	—	—	0.31	0.04	0.66
Perceived pleasure	—	—	—	0.39	—	—	—	—	—	—	—
Usage of the system	—	—	—	—	0.12	0.60	—	—	—	—	—
Attitude to utilization	—	—	—	—	—	0.23	—	—	—	0.59	—

$$\begin{aligned}\text{Intuition of use} = {} & 0.679 + (0.069 \times \text{individual norms}) \\ & + (0.067 \times \text{image}) - (0.045 \times \text{perceived risk}) \\ & + (0.356 \times \text{usage of the system}) \\ & + (0.080 \times \text{attitude to utilization}) \\ & + (0.238 \times \text{perceived pleasure})\end{aligned}$$

$$\begin{aligned}\text{Perceived pleasure} = {} & 2.645 \\ & + (0.281 \times \text{Pragmatic shopping orientation})\end{aligned}$$

$$\begin{aligned}\text{Satisfaction} = {} & 0.725 + (0.625 \times \text{Intuition of use}) \\ & + (0.239 \times \text{e-commerce service quality})\end{aligned}$$

To summarize, usage of the system is strongly dependent on personal innovativeness and website quality. This means that it is within the rights of PSCs to increase the purchasing level of their website quality. Additionally, intuition of use is affiliated with individual norms, image, perceived risk, usage of the system, attitude to utilization, and perceived pleasure, which is hardly influenced by personal views. Perceived pleasure is affected by utilitarian and hedonic shopping orientation, which can be indirectly controlled by PSCs. Finally, satisfaction is imposed from intuition of use and e-commerce website quality, which can be inspected by PSCs (Figure 8.3).

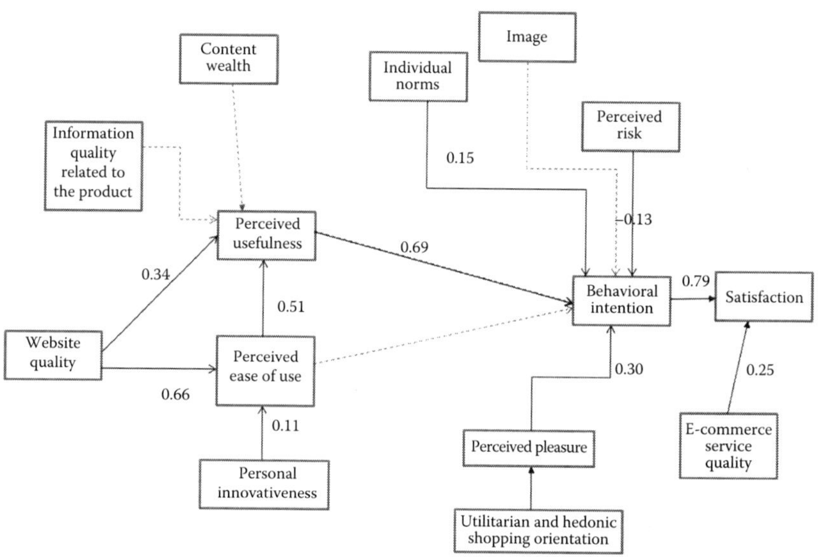

Figure 8.3 Accepted H_i hypotheses (continuous line) and rejected H_i hypotheses (dotted line) and their coefficients.

8.4 Future Research Directions

This research can be expanded with structural equation modeling in order to examine the relationship between independent variables. Moreover, another survey can be conducted to compare the effectiveness of these two methods.

8.5 Conclusion

This study aims to demonstrate the significant factors that affect the purchasing behavior of private shopping end users in e-commerce. To determine the expectations of the customers, multiple regression analysis is integrated with the psychological attributes such as individual norms, image, perceived risk, usage of the system, attitude to utilization, and perceived pleasure in accordance with an effective representation of the TAM. From the results of this research, the intention to use PCS business models can be analyzed by companies. Moreover, PSC users' expectations from PSC websites can be specified, which will enable PCSs to assess the customers' priorities from their purchasing behavior. Additionally, they can actualize their intention of sales behavior through the findings of this study. Thus, they can also analyze customer satisfaction at the end of their actual purchase.

References

Aboelmaged, M.G. (2010). Predicting e-procurement adoption in a developing country: An empirical integration of technology acceptance model and theory of planned behavior, *Industrial Management & Data Systems*, 110, 392–414.

Anumba, C.J. and Ruikar, K. (2002). Electronic commerce in construction-trends and prospects, *Automation in Construction*, 11, 265–275.

Atchariyachanvanich, K., Okada, H., and Sonehara, N. (2007). Theoretical model of purchase and repurchase in internet shopping: Evidence from Japanese online customers. In *ICEC'07*, Minneapolis, MN.

Browne, M.W. and Cudeck, R. (1993). *Testing Structural Equation Models*. Newbury Park, CA: Sage Publications.

Chen, D.N., Jeng, B., Lee, W.P., and Chuang, C.H. (2008). An agent-based model for consumer-to-business electronic commerce, *Expert Systems with Applications*, 34, 469–481.

Choudhury, V. and Hartzel, K.S. (1998). Uses and consequences of electronic markets: An empirical investigation in the aircraft parts industry, *MIS Quarterly*, 22, 471–503.

Davis, F.D. (1989). A technology acceptance model for empirically testing new end user information systems: Theory and results. Doctoral Dissertation, MIT Sloan School of Management, Cambridge, MA.

Hawes, J.M. and Lumpkin, J.R. (1986). Perceived risk and the selection of a retail patronage mode, *Journal of the Academy of Marketing Science*, 14, 37–42.

Ho, S.C., Kauffman, R.J., and Liang, T.P. (2007). A growth theory perspective on B2C e-commerce growth in Europe: An exploratory study, *Electronic Commerce Research and Applications*, 6, 237–259.

Hsia, T.L., Wu, J.H., and Li, E.Y. (2008). The e-commerce value matrix and use case model: A goal-driven methodology for eliciting B2C application requirements, *Journal of Information Management*, 45, 321–330.

Jiang, X. and Qian, X. (2009). Study on intelligent e-shopping system based on data mining, School of Electronic Information & Electrical Engineering, Changzhou Institute of Technology CZU, Chanzhou, China.

Jun, M., Yang, Z., and Kim, D. (2004). Customers' perceptions of online retailing service quality and their satisfaction, *International Journal of Quality & Reliability Management*, 21, 817–840.

Lee, Y., Kozar, K.A., and Larsen, K.R.T. (2003). The technology acceptance model: past, present, and future, *Communications of the Association for Information Systems*, 12, 752–780.

Legris, P.J., Ingham, P., and Collerette, P. (2003). Why do people use information technology? A critical review of the technology acceptance model, *Information and Management*, 40, 191–204.

Nath, R., Akmanligil, M., Hjelm, K., Sakaguchi, T., and Schultz, M. (1998). Electronic commerce and internet: Issues, problems and perspectives, *International Journal of Information Management*, 18, 91–101.

Navarro, J.G.C., Jimenez, D., and Conesa, E.A.M. (2007). Implementing e-business through organizational learning: An empirical investigation in SMEs, *International Journal of Information Management*, 27, 173–186.

Numberger, S. and Rennhak C. (2005). The future of B2C e-commerce, *Electronic Markets*, 15, 269–282.

Peppers, D. and Rogers, M. (2001). *One to One B2B: Customer Development Strategies for the Business-to-Business World*. New York: Doubleday.

Quaddus, M. and Achjari, D. (2005). A model for electronic commerce success, *Telecommunications Policy*, 29, 127–152.

Wang, Y.S. (2002). The adoption of electronic tax filing systems: An empirical study, *Government Information Quarterly*, 20, 333–352.

9

Traffic Signal Optimization

Challenges, Models, and Applications

MAHMUT ALİ GÖKÇE,
ERDİNÇ ÖNER, AND GÜL IŞIK

Contents

9.1 Introduction

Traffic congestion is one of the most important problems of urban life. Urban traffic system composed of vehicles, pedestrians, traffic lights, and traffic network structure results in a complex problem to be solved (Salimifard and Ansari, 2013). As much as new roads are being built, because the (rate of expansion of) demand for new capacity on roads and other transportation systems exceed the (rate of expansion of) supply, congestion continues to cause an increasing loss of valuable time and resources in urban areas.

Different approaches may be used to minimize the traffic congestion problem in urban traffic systems. Improving public transportation and intelligent transportation system applications are some of the main approaches being used. One other such approach is the effective use of traffic signals. Traffic signal systems are used to control and regulate traffic flow for pedestrians and vehicles at road intersections or other locations, where heavy flow of traffic is present (Roess et al., 2004). Traffic congestion is heavily a flow problem. Therefore, it can be managed depending on how the combination of various signals on the way is regulated. Researchers often agree that the correct staging of traffic signals can help to reduce traffic congestion by improving the flow of the vehicles in urban areas (Garcia-Nieto et al., 2013).

This chapter is on traffic signal optimization. In next section, we discuss the challenges associated with the traffic signal optimization problem. Under Section 9.3, we present a review of models to manage different types of traffic signal optimization problems with examples of applications.

9.2 Challenges

Challenges of the traffic signal optimization problem can be summarized under the following three headings: coordination, scale, and measurement.

9.2.1 Coordination

The traffic signal operations are generally designed for individual locations (single intersections). However, to improve and regulate the traffic flow in urban traffic system, the coordination of traffic signals or the whole traffic network system should be taken into consideration. The effects of the traffic signal operations at any intersection on the downstream/upstream have to be considered. The design of the traffic signal operations in one intersection may improve the traffic flow there, whereas in the other downstream/upstream intersection(s), it might degrade the traffic flow.

The capacity of the intersections and the area available at the traffic signals are other important characteristics, which have to be considered, for the urban traffic system. The higher number of intersections in urban areas resulting in shorter distances between the intersections may not have enough capacity to hold high number of vehicles at the traffic signals, which would result in spillback queues to arterial roads.

In addition to the capacity of the intersections to improve the urban traffic flow, undisrupted traffic flow should be provided to reduce urban traffic congestion. The short distances between the intersections may result in a stop & go traffic flow condition. Green waves and coordinated traffic signal optimization approaches are used to overcome the stop&go situation (Pengdi et al., 2012).

9.2.2 Scale

The optimization of the traffic signal operations in the urban traffic system is a complex problem since an intersection consists of a number of approaches and the crossing area (Dotoli et al., 2006). The traffic signal light switches to red and green requires the introduction of the discrete variable for the traffic signal timing optimization problem, which makes the problem combinatorial, and considering the whole urban traffic system for the optimization, the problem becomes very large very quickly (Papageorgiou et al., 2003). Another difficulty in solving the traffic signal timing optimization problem is the unpredictable disturbances (incidents, illegal parking, intersection blocking, etc.) on the traffic flow, which also introduce the stochasticity to the urban traffic problem.

9.2.3 Measurement

The prerequisite of improving any system is to be able to measure the effects of the changes made to the system. If one wants to optimize traffic signal operations in an urban traffic system, for measuring the effects, ideally real-time data are required. Due to the nature of the traffic system, the data are variable during the day, week, month, or the year. The number of vehicles, percentages of different types of vehicles in the urban traffic system, the distance and location information of the intersections, the number of pedestrians, and entry and exit points of the vehicles and pedestrians are required to solve the traffic signal optimization problem optimally. With the help of the technological advancements, it is becoming easier to collect traffic data. Inductive loop detectors, radar, microwave, and ultrasonic sensors to infrared cameras and in-vehicle GPS/GSM receivers/transmitters (*floating car data*) are few of the sensor technologies used to collect traffic data (Van Lint and Hoogendoorn, 2010). The use of real-time traffic data to solve the traffic signal optimization problem also requires the real-time evaluation and analysis of the data. Developing models for the analysis of these large amounts of data is a great challenge.

9.3 Models

Traffic signal optimization problem is studied either for a single intersection or for a network of intersections. A single intersection is by nature isolated and relatively easier to study. Although network of intersections are much more realistic than a single intersection, modeling and solving those models become harder very quickly. Sometimes, a single intersection is so isolated that one can study and arrange its signal timings to get significant improvement in the traffic flow.

There are mainly three types of plan for traffic signal control. These are fixed-time, semiactuated, and fully actuated types.

Fixed-time plans, also called pretimed plans, are the most basic for which each phase of the signal lasts for a specific duration of time before changing into the next phase. Fixed-time plans usually utilize historical data to determine the timings. Although for this type of plan, the timings are independent of the current traffic flow, multiple timing settings can be used for different times of the day. The advantage of fixed-time plan is that they are relatively cheap to implement.

Actuated plans, semi or fully, are traffic responsive but require significantly more investment to implement. Vehicle sensors and detectors need to be installed at many intersections as well as an algorithm to manage the real-time data collected. Some of the simpler actuated plans choose the best fixed-time plan among the stored in the system that best fits the real-time data collected at that time.

Because traffic light signal timing is one of the cheapest and most effective methods of reducing traffic congestion in metropolitan roads and networks (Spall and Chin, 1997), the problem of signal timing optimization has been studied with a variety of methods. We present here some of these methods grouped by the approach used by the method.

9.3.1 Analytical Optimization Models

Signal timing optimization requires measurement of performance for a particular setting of the signal timing. To develop an analytical model, which can take into account the different vehicles from different directions, at nondeterministic rates is challenging. Therefore, there are, to the best knowledge of the authors, relatively fewer analytical models for signal timing optimization models in literature.

For a single intersection, a fixed-time strategy can be either stage-based or phase-based. Stage-based strategy determines the optimal cycle and split times. Phase-based fixed-time strategy also determines stage specifications, which includes the options (turning left, going straight, etc.) a vehicle has at an intersection.

One of the earliest and well-known stage-based fixed-time strategies was SIGSET (Allsop, 1971). SIGSET's objective function was a nonlinear delay function derived by Webster (1958). SIGSET used m linear constraints on the capacity of stage specifications, which resulted in a linearly constrained nonlinear programming problem. SIGCAP was also developed by Allsop, which aimed to maximize intersection's capacity (Allsop, 1976). SIGCAP, however, was a linear programming problem.

Improta and Canteralla (1984) solved a similar single intersection signal optimization problem but included stage specifications (phase-based fixed-time strategy). Their approach determined split and cycle time, as well as stage specifications to optimize total delay or system capacity. To determine stage specifications, they had to add binary variables, which made their problem a mixed-integer linear programming (MILP) problem with increased difficulty.

For a network of intersections, one of the earliest MILP models is developed by Little in 1966 by the name MAXBAND (Little, 1966). Little developed a MILP for an n intersection of a two-way arterial, for which split and cycle times are assumed to be given. His model determined the optimal offset based on maximizing the number of vehicles that can travel a given range without stopping. Later, MAXBAND was transformed into a portable Fortran code, which was able to handle three-artery networks with up to

17 signals (Little et al., 1981). Chaudhary et al. (1991) reduced computational requirements of MAXBAND. Stamatiadas and Gartner (1996) extended MAXBAND so that it became applicable to networks of arterials.

9.3.2 Heuristic Optimization Models

Urban traffic analysis and control is a problem, which has complexity making it difficult to analyze with traditional analytical methods (Tartaro et al., 2001).

Even Webster, in his work from 1958, states "Since a theoretical calculation of delay is very complex and direct observation of delay on the road is complicated by uncontrollable variations, it was decided to use a method whereby the events on the road are reproduced in the laboratory by means of some machine which simulates behavior of traffic...," suggesting the use of simulation for traffic problems.

Hewage and Ruwanpura (2004) stated that computer simulation can be useful to analyze traffic flow patterns and signal light timings.

Urban Traffic Control System was the first traffic simulation software developed under the direction of Federal Highway Administration (FHWA) and later named NETSIM, short for Network Simulation. FREESIM (abbreviated from Freeway Simulation) was an enhanced version of NETSIM, which could handle more complex freeway geometrics and provide a more realistic representation of traffic on a freeway. In 1998, NETSIM and FREESIM were combined and offered to public under the new name CORSIM.

There are two main approaches in traffic modeling using simulation, which are microscopic and macroscopic approaches. In a microscopic model, each car is simulated individually whereby dynamic variables of the models represent microscopic properties like the position and velocity of single vehicles. Macroscopic models take more an aggregated approach, where it is assumed that traffic flows as a whole are comparable to fluid streams. Therefore, in this case, one is more interested in traffic flow characteristics like density, mean speed of traffic flow, etc.

TRANSYT-7F is a macroscopic traffic simulation model, which also does optimization using genetic algorithms (GA) and a hill climbing method. It was originally developed in the

United Kingdom by Transport and Road Research Laboratory. It was later adapted by FHWA, thus acquiring *7F* after version 7. TRANSYT-7F considers platoons of vehicles instead of individual vehicles but simulates the flow in small time increments, which allows more detailed representation than most other macroscopic models. TRANSYT-7F allows the user to choose from either a GA or a hill climbing method to optimize signal timings (*TRANSYT-7F's User's Manual*, 1998).

Synchro evaluates a series of cycle length while applying a heuristic method to determine green splits to optimize the four signal timing parameters (Synchro, 2014). During these evaluations, it also conducts an exhaustive search for left-turn phase position and a quasiexhaustive search for offsets. Synchro uses percentile of traffic flow as the optimization criterion.

But TRANSYT-7F and Synchro are unable to fully consider the important aspects of traffic behaviors due to the nature of the macroscopic simulation model. For example, lane-changing behavior and vehicle interactions are not considered in the macroscopic models in a realistic manner.

Although there are studies with macroscopic models, there is a growing recognition of the usefulness of stochastic microscopic simulation models (Lindgren and Tantiyanugulchai, 2003).

Microscopic simulation models allow realistic scenarios to be tested under real-world conditions and also provide network-wide performance measures like travel times, delays, emissions, etc. (White, 2001). Microscopic simulation models are becoming a more accepted tool for signal timing and capacity studies (Park et al., 2003).

CORSIM was among the earliest microscopic simulation tools developed. Today VISSIM, SUMO, PARAMICS, and SIMTRAFFIC are just some of the microscopic models available. Park and Yun (2003) compared various microscopic simulation models in terms of computation time and capability of modeling a coordinated actuated signal system. An earlier but a much wider review of microscopic models is given by Algers et al. (1997) in the SMARTEST project.

Among the heuristic optimization methods, especially use of GA is dominant for determining signal timing. Foy et al. (1992) proposed a GA to determine signal timing for a two-phase system.

Hadi and Wallace (1993) developed a GA to be used in combination with the TRANSYT-7F optimization routine to determine signal timing and phasing. Clement and Taylor (1994) developed a GA and a knowledge-based system for dynamic traffic signal control. Abu-Lebdeh and Benekohal (1997, 1998) applied GAs to oversaturated arterials for traffic control and queue management. Park et al. (2001) applied GA with CORSIM microscopic traffic simulation package. Park and Schneeberger (2003) applied GA with VISSIM microscopic traffic simulation package.

The list of GA-based signal timing optimizations that are based on single-objective searches are long and cannot be covered here comprehensively (see the literature review in Stevanovic et al., 2007). But there are also studies that implemented an evolutionary multiobjective optimization to retime traffic signals. Sun et al. (2003) applied NSGA-II to optimize delay and stops for an isolated intersection under two-phase control. NSGA-II obtained a close approximation to the Pareto set while using an analytical formula to determine delay and stops. Abbas et al. (2007) applied NSGA-II to a small three-signal network while choosing signal settings from a predetermined set obtained from a single objective optimization at different cycle lengths. Thus, the multiobjective optimizer considered only variations in cycle length, simplifying the exercise considerably. Kesur (2010) investigated and suggested a multiobjective optimization when there are numerous optimization variables.

Use of other metaheuristic methods to optimize signal timing has been limited. Chen and Xu (2006) applied particle swarm optimization (PSO) for training a fuzzy logic controller located at intersections with the aim of determining green light timings. They used a simple network with two basic junctions to test their model. More recently, Peng et al. (2009) developed a PSO algorithm for a restrictive one-way road with two intersections with custom microscopic traffic flow model. Finally, Garcia-Nieto et al. (2013) proposed a PSO algorithm to optimize traffic light cycle programs using a microscopic traffic simulator, SUMO.

Işık et al. (2013) proposed a PSO algorithm to optimize signal timings using VISSIM microscopic traffic simulation as an evaluation function and presented a real-life application for a 28-signal head timing optimization in a roundabout in Izmir, Turkey. PSO model goes through different cycle times and finds best split times for each signal head for each cycle time. Results from the experimentation show that their proposed model provides a 10% increase in the number of vehicles passing through the roundabout and a 56% decrease in the average delay through the roundabout compared to the current system settings.

There are also some studies that utilize Neural Networks, Cellular Automata, fuzzy control, and fuzzy-neuro methods for urban traffic signal control, but they diverse quite a bit in the settings that they consider and, for that reason, are not included here.

9.4 Conclusion

Traffic congestion in urban areas is a growing problem. To manage and regulate traffic in these urban settings, traffic lights are used. The cheapest and most effective method of providing any improvement for traffic congestion goes through better managing the traffic signal timing. Signal timing improvement requires almost no hardware investment and can easily be implemented after being carefully developed in controlled environments. This chapter is on traffic signal optimization. The challenges associated with traffic signal optimization problem are discussed. Also, an extensive review of different models and methods used to solve different types of traffic signal optimization problems is provided. Although authors make no claim as to this chapter being a definitive or a complete review, it is a good resource for anyone interested in traffic signal optimization to learn about the important challenges and models of the subject matter.

References

Abbas, M.M., H.A. Rakha, and P. Li. 2007. Multi-objective strategies for timing signal systems under oversaturated conditions. *Proceedings of the 18th IASTED International Conference*, Montreal, Quebec, Canada, pp. 580–585.

Abu-Lebdeh, G. and R.F. Benekohal. 1997. Development of a traffic control and queue management procedure for oversaturated arterials. *Proceedings of the 76th Transportation Research Board Annual Meeting*, Washington, DC.

Abu-Lebdeh, G. and R.F. Benekohal. 1998. Evaluation of dynamic signal coordination and queue management strategies for oversaturated arterials. *Proceedings of the 76th Transportation Research Board Annual Meeting*, Washington, DC.

Algers, et al. 1997. SMARTEST final report. http://www.its.leeds.ac.uk/projects/smartest/finrep.PDF (last accessed June 12, 2014).

Allsop, R.B. 1971. SIGSET: A computer program for calculating traffic capacity of signal-controlled road junctions. *Traffic Engineering & Control*, 12, 58–60.

Allsop, R.B. 1976. SIGCAP: A computer program for assessing the traffic capacity of signal-controlled road junctions. *Traffic Engineering & Control*, 17, 338–341.

Chaudhary, N.A., A. Pinnoi, and C. Messer. 1991. Proposed enhancements to MAXBAND-86 Program. Transportation Research Record 1324, pp. 98–104.

Chen, J. and L. Xu. 2006. Road-junction traffic signal timing optimization by an adaptive particle swarm algorithm. *Proceedings of the Ninth International Conference on Control, Automation, Robotics and Vision*, Vols. 1–5, pp. 1–7.

Clement, S.J. and M.A. Taylor. 1994. The application of genetic algorithms and knowledge-based systems to dynamic traffic signal control. *Proceedings of the Second International Symposium on Highway Capacity*, Vol. 1, pp. 193–202.

Dotoli, M., M. Pia Fanti, and C. Meloni. 2006. A signal timing plan formulation for urban traffic control. *Control Engineering Practice*, 14(11), 1297–1311.

Foy, M., R.F. Benekohal, and D.E. Goldberg. 1992. Signal timing determination using genetic algorithms. Transportation Research Record 1365, pp. 108–115.

Garcia-Nieto, J., A.C. Olivera, E. Alba. 2013. Optimal cycle program of traffic lights with particle swarm optimization. *IEEE Transactions on Evolutionary Computation*, 17(6), 823–839.

Hadi, M.A. and C.E. Wallace. 1993. Hybrid genetic algorithm to optimize signal phase and timing. Transportation Research Record 1421, pp. 104–112.

Hewage, K.N. and J.Y. Ruwanpura. 2004. Optimization of traffic signal light timing using simulation. *Proceedings of the Winter Simulation Conference*, Vol. 2, pp. 1428–1433.

Improta G. and G.E. Cantarella. 1984. Control systems design for an individual signalised junction. *Transportation Research B*, 18, 147–167.

Işık, G., E. Öner, and M.A. Gökçe. 2013. Traffic signal timing optimization for a signalized roundabout in Izmir. *The International IIE (Institute of Industrial Engineers) Conference/YAEM*, June 26–28, 2013, İstanbul, Turkey.

Kesur, K.B. 2010. Generating more equitable traffic signal timing plans. Transportation Research Record 2192, pp. 108–115.

Lindgren, R.V. and S. Tantiyanugulchai. 2003. Microscopic simulation of traffic at a suburban interchange. *2003 Annual Meeting of the Institute of Transportation Engineers*, Seattle, WA.

Little, J.D.C. 1966. The synchronisation of traffic signals by mixed-integer-linear-programming. *Operations Research*, 14, 568–594.

Little, J.D.C., M.D. Kelson, and N.H. Gartner. 1981. MAXBAND: A program for setting signals on arteries and triangular networks. Transportation Research Record 795, pp. 40–46.

Papageorgiou, M., C. Diakaki, V. Dinopoulou, A. Kotsialos, and Y. Wang. 2003. Review of road traffic control strategies. *Proceedings of the IEEE*, 91(12), 2043–2067.

Park, B., N.M. Rouphail, and J. Sacks. 2001. Assessment of a stochastic signal optimization method using microsimulation. Transportation Research Record 1748, pp. 40–45.

Park, B. and J.D. Schneeberger. 2003. Microscopic simulation model calibration and validation: A case study of VISSIM for a coordinated actuated signal system. Transportation Research Record 1856, pp. 185–192.

Park, B., I. Yun, and K. Choi. 2003. Evaluation of microscopic simulation programs for coordinated signal system. In *13th ITS America's Annual Meeting*, Minneapolis, MN, May 9–22.

Peng, L., M.-H. Wang, J.-P. Du, and G. Luo. 2009. Isolation niches particle swarm optimization applied to traffic lights controlling. *Proceedings of 48th IEEE Conference on Decision Control/28th Chinese Control*, pp. 3318–3322.

Pengdi, D., N. Muhan, W. Zhuo, Z. Zundong, and D. Honghui. 2012. Traffic signal coordinated control optimization: A case study. *24th Chinese Control and Decision Conference (CCDC)*, Taiyuan, China, May 23–25, 2012, pp. 827–831.

Roess, R.P., E.S. Prassas, and W.R. McShane. 2004. *Traffic Engineering*, 3rd edn. Upper Saddle River, NJ: Pearson/Prentice Hall.

Salimifard, K. and M. Ansari. 2013. Modeling and simulation of urban traffic signals. *International Journal of Modeling and Optimization*, 3(2), 172–175.

Spall, J.C. and D.C. Chin. 1997. Traffic-responsive signal timing for system-wide traffic control. Transportation Research—C, Vol. 5, pp. 153–163.

Stamatiadisand C. and N.H. Gartner. 1996. MULTIBAND96: A Program for Variable Bandwidth Progression Optimization of Multiarterial Traffic Networks. Transportation Research Record, No. 1554, pp. 917.

Stevanovic, A., P.T. Martin, and J. Stevanovic. 2007. VISGAOST: VISSIM-based genetic algorithm optimization of signal timings. Transportation Research Record 2035, pp. 59–68.

Sun, D., R.F. Benekohal, and S.T. Waller. 2003. Multi-objective traffic signal optimization using non-dominated sorting genetic algorithm. *IEEE Intelligent Vehicles Symposium*, Piscataway, NJ, pp. 198–203.

Synchro. http://208.131.129.243/wp-content/uploads/2013/08/SignalTiming Background.pdf (last accessed January 30, 2014).

Tartaro M.L., C. Toress, and G. Wainer. 2001. Defining models of urban traffic using the TSC tool. *Proceedings of the Winter Simulation Conference*, pp. 1056–1063.

TRANSYT-7F's User's Manual. Transportation Research Center, University of Florida, Gainesville, FL, March 1998.

Van Lint, J.W.C. and S.P. Hoogendoorn. 2010. A robust and efficient method for fusing heterogeneous data from traffic sensors on freeways. *Computer-Aided Civil and Infrastructure Engineering*, 25, 596–612.

Webster, F.V. 1958. Traffic signal settings. Road Research Technical Paper No. 39, Research Laboratory, London, U.K.

White, T. 2001. General overview of Simulation Models. *49th Annual Meeting of Southern District Institute of Transportation Engineers*, Williamsburg, VA.

10

Comparative Financial Efficiency Analysis for Turkish Banking Sector

A. ARGUN KARACEBEY
AND FAZIL GÖKGÖZ

Contents

10.1 Introduction

The financial decision makers desire to measure the efficiency level of a DMU by considering the positive and negative conditions. In this framework, efficiency analysis becomes a vital instrument for an enterprise operating under global competitive market conditions.

CCR model proposed a widely used efficiency measurement technique that is known as the DEA (Charnes et al. 1978). It is advantageous since the DEA technique does not need a functional relationship between inputs and outputs (Charnes et al. 1978, Choi and Murthi 2001).

However, the DEA has become popular in evaluating technical and pure efficiencies since the method easily processes multiple outputs without requiring the input price data (Ruggiero 2001). There are numerous financial efficiency analyses applied on the banking sector—some empirical studies such as Baurer (1993), Berger and Humprey (1997), and Berger et al. (1993) had carried out frontier efficiency analyses for the US banks, while other studies, as Carbo et al. (2002) and Conceicao et al. (2007), performed efficiency measurements for the banks in the European Union. Thus, similar efficiency analyses have been carried out for the Turkish banking system and introduced particularly in the studies of Çingi and Tarım (2000), Ertuğrul and Zaim (1999), Karacabey (2002), Kasman (2002), Mercan and Yolalan (2000), and Zaim (1995).

Due to the financial crises encountered in the Turkish economy in 2000 and 2001, the financial efficiency estimations for the Turkish banking system have become crucial in evaluating the vital components of the financial system of the country.

The aim of this chapter is to measure the financial efficiency of the Turkish COMs and INVs for the 2010–2011 period and compare the technical efficiency (TE), pure technical efficiency (PTE), and scale efficiency (SE) of these DMUs. In this framework, the CCR and BCC versions of the DEA models were applied to the COMs and INVs of Turkey so as to find out the financial efficiency score levels.

The remainder of the chapter is organized as follows: Section 10.2 briefly explains the characteristics of the Turkish banking and finance sector, while Section 10.3 gives a detailed account of the DEA model. Section 10.4 describes the methodology with regard to the data used and presents the results. Section 10.5 provides the concluding remarks.

10.2 Current Status of Banking and Finance Sector in Turkey

Crisis experience of Turkey emphasizes the supplementary relation between macroeconomic and financial stability and structural reforms (BRSA 2010). Weak banking sector system was one of the main causes for Turkey's 2000 and 2001 crises that adversely affected the Turkish economy. With these financial crises, the Turkish banking sector underwent a restructuring process, during which 14 banks were transferred to the Saving Deposits Insurance Fund (SDIF) between 2000 and 2003, as they were not able to meet their liabilities (SDIF 2003).

This process was beneficial for both solving the financial problems of the Turkish banking sector and for aligning the banking legislation with the international qualified implementations. In 2002, the Turkish economy began to recover, experiencing 6.8% annual growth in average between 2002 and 2007. High economic growth strengthened the banking sector, which in turn contributed positively to the economic growth.

The number of the banks decreased to 50 from 59 as of December 2006 by the impact of the consolidation experienced between 2002 and 2006. Yet, the Turkish banks widened their branch structure and increased their personnel in accordance with the accelerated economic growth between 2002 and 2007 (BRSA 2006, 2007).

The global financial crisis, which broke out in 2008, affected all countries' economic stability, and subsequently global economic growth has decelerated. Shrunk by 4.1% in the 2008–2009 period, the Turkish economy recovered in 2010 and grew with an annual average of 9% in the period between 2010 and 2011. Experiencing relatively limited adverse effects of the global financial crisis, the Turkish banking sector is continuing to expand its branch structure. However, following the crisis, the growth rate in the number of branches has diminished consequently since the banks had to reevaluate their branching predictions. Moreover, it is considered that the proliferation of call centers and Internet banking has contributed to this trend (BRSA 2006).

The banking sector improved its operational efficiency through intensifying the use of technology. Alternative distribution channels such as online banking, ATM, and call centers increased in parallel with the technologic developments in electronic environment of the banking sector. The volume of the financial transactions made a progress both by offering a widespread accessibility and ensuring saving time and costs (BRSA 2011).

However, during the same period, the growth rate of the personnel number also decreased due to the global financial crisis. It is thought that the contribution of the financial institutions to total employment is quite low. In the period between 2008 and 2011, the banking sector's contribution to employment was 3.1% in average (BRSA 2011).

Although the nonbanking financial sector has grown in number and size in Turkey, banks still dominate the sector. Set alongside banks, the Turkish financial sector largely comprises insurance and private pension companies. Nonbanking financial institutions like factoring, leasing, consumer financing companies, and intermediary institutions also operate in the sector.

Asset sizes of the main financial services subsectors in Turkey as they stood at the end of 2011 are shown in Table 10.1. In the period between 2002 and 2011, the asset size of the Turkish financial sector has substantially increased. In this period, the banking sector assets have been risen to 1.217.6 billion TL from 212.7 TL as of 2011, implying approximately a fivefold increase (BRSA 2009, 2011).

Profitability of the banking sector fluctuated in the last decade due to various factors such as changes in profits and growth of assets. Despite the global financial crisis, the sector succeeded in maintaining its precrisis approx. 2% return on assets ratio in the 2008–2011 period (TBB 2011).

High-quality capital structure of the banking sector is a protective indicator of possible financial or macroeconomic fluctuations (BRSA 2011). The banking sector's strong capital structure contributes to the maintenance of the economic growth at a sustainable level. On the other hand, the capital adequacy ratio tends to decrease since 2003 due to the relative high increases in assets. Stood at 30.9% at the end of 2003, capital adequacy ratio of the banking sector declined until 2008 to the level of 18%. Decreasing asset size due to the global financial crisis caused the ratio to jump to 20.6% in 2009. In the 2010–2011 period, it continued to decrease and realized as 16.5% at the end of 2009. However, the capital adequacy ratio of the banking sector is still above the target ratio, which is 12% (Treasury 2013).

10.3 Literature

10.3.1 DEA Models

The DEA is a nonparametric and a linear programming technique that has been used to compare the TE of relatively homogeneous

Table 10.1 Asset Size of the Main Financial Services Subsectors in Turkey

BILLION TL	2002	2003	2004	2005	2006	2007	2008	2009	2010	2011
Banks	212.7	249.7	306.4	406.9	499.7	581.6	732.5	834.0	1006.0	1217.6
Financial leasing	3.8	5	6.7	6.1	10	13.7	17.1	14.6	15.7	18.6
Factoring	2.1	2.9	4.1	5.3	6.3	7.4	7.8	10.4	14.5	15.7
Consumer financing	0.5	0.8	1.5	2.5	3.4	3.9	4.7	4.5	6.0	8.9
Insurance	5.4	7.5	9.8	14.4	17.4	22.1	26.5	31.8	35.1	39.9
Secur. Int. Inst.	1	1.3	1	2.6	2.7	3.8	4.2	5.2	7.5	9.6

Source: Banking Regulation and Supervision Authority of Turkey, Financial market reports, 2009, 2011.

sets of DMUs. The theoretical consideration of TE has existed in the economic literature since Koopmans (1951) defined TE as a feasible input/output vector where it is technologically impossible to increase any output without simultaneously (Ruggiero 2000).

Farrell performed one of the earliest studies regarding the efficiency measurement for the homogeneous DMUs. However, Farrell (1957) perceived the efficiency as in technical and allocative basis, and determined the efficiencies under the framework of one output and multiple inputs. Besides, the DEA has the advantage of having multiple output structure for the efficiency analysis. The DEA can be stated as a data-oriented, nonparametric,* and a linear programming technique that has been used to compare the TE of relatively homogeneous sets of DMUs.

Thus, the DEA has been applied successfully to finance sector as well. Murthi et al. (1997) affirm that the DEA does not need a theoretical model as measurement benchmarks. Moreover, the DEA may address the problem of endogeneity of transaction costs by considering the transaction costs such as expense ratio and turnover, and the aforementioned model is flexible and may evaluate the performance on a number of outputs and inputs simultaneously. The DEA method also facilitates the observation of the marginal contribution of each input in affecting returns. As a consequence of these advantages, the method permits the analysis of the performance level of a particular DMU in comparison to the efficiency (Gökgöz 2009a,b, 2010). The efficiency score of the DEA can be defined as the ratio between a weighted sum of outputs and a weighted sum of inputs. The objective function of the DEA for n DMUs consuming k inputs and producing m outputs is as follows (Gökgöz 2009a,b, 2010):

$$\text{Max}\frac{u'y_i}{v'x_i} \tag{10.1}$$

* Parametric and nonparametric approaches are the well-known quantitative techniques that can be classified as the efficient Frontier Approach. Namely, the Stochastic Frontier Approach, Distribution Free Approach, and Thick Frontier Approach are the parametric efficiency measurement techniques, whereas the Free Disposal Hull and Data Envelopment Analysis (DEA) are considered the nonparametric approaches.

where

u' is the output weight vector ($m \times 1$)
y_i is the amount of output produced by DMU_i
v' is the input weight vector ($k \times 1$)
x_i is the amount of input utilized by DMU_i

The efficiency score lies between 0 and 1 for input-oriented model, while output-oriented model efficiency score ranges between 1 and ∞. For both models, the DMUs having efficiency score as 1 is considered efficient.

Efficiency results of the DEA give efficient and inefficient DMUs according to constant return to scale (CRS) and variable return to scale (VRS) assumptions. However, these results also reveal slack inefficiency levels for the inefficient observations.

There are two basic DEA models on the basis of orientation. The output-oriented model assumes the capacity of a DMU to reach the maximum production level (output) under available inputs. The input-oriented model refers the ability to produce the same capacity of production with the minimum input level (Cooper et al. 2000, 2006).

TE scores of DMUs are measured by the CCR model that was introduced by Charnes et al. (1978). The model* depends upon CRS† assumption (Fandel 2003). Input-oriented CCR model is explained as in the formulation given as follows (Cooper et al. 2000, 2006):

$$\text{Max } u'y_i \tag{10.2}$$

s.t.

$$v'x_i = 1 \tag{10.3}$$

$$u'y_i - v'x_i \leq 0 \tag{10.4}$$

$$u, -v \geq 0$$

* Fandel states that a DMU will demonstrate technical efficiency, if TE equals to "1" according to the CCR model. Besides, if TE < 1, then it shows to what extent a particular DMU should minimize inputs for producing its level outputs as efficient as technically efficient DMUs.

† CRS assumes a proportional relationship between the increases in the inputs and the outputs.

where

u' is the output weight vector ($m \times 1$)
y_i is the amount of output produced by DMU_i
v' is the input weight vector ($k \times 1$)
x_i is the amount of input utilized by DMU_i
u is the output weights
v is the input weights
($i = 1, \ldots, n$)

To determine TE scores for DMUs, input-oriented model has n optimizations, and all individual DMUs choose input and output weights that maximize TE scores. Alternatively, the BCC model was proposed to analyze PTE and SE of DMUs by considering the VRS* assumption (Banker et al. 1984). Banker et al. (1984) affirm that the convexity constraint† is an additional restriction for the BCC model structure (Barrientos and Boussofiane 2005). The input-oriented BCC model is formulated as follows (Cooper et al. 2000, 2006):

$$\text{Min}_{\theta,\lambda}\theta \tag{10.5}$$

s.t.

$$\theta x_i - X\lambda \geq 0 \tag{10.6}$$

$$Y\lambda \geq y_i \tag{10.7}$$

$$\sum \lambda = 1 \tag{10.8}$$

$$\lambda \geq 0 \tag{10.9}$$

where

θ is the PTE score (scale ratio)
y_i is the amount of output produced by DMU_i
x_i is the amount of input utilized by DMU_i

---

* DMUs are sometimes unable to operate in optimal scales. The proportional variation at inputs may result in different proportional variation of outputs.

† Equation 10.8 refers to the convexity constraint at which the sum of the multipliers should add to "1."

λ is the $n \times 1$ sized vector
Y is the $n \times s$ sized output matrix
X is the $n \times m$ sized input matrix
$(i = 1, \ldots, n)$

Within the framework of the BCC model, total TE is composed of PTE and SE. However, in order to calculate the SE, TE score should be divided by PTE score for each DMU. The SE is formulated for DMUs as follows (Cooper et al. 2000, 2006):

$$\text{SE} = \frac{\text{TE}_{\text{CCR}}}{\text{PTE}_{\text{BCC}}} \tag{10.10}$$

As the difference between TE and PTE scores increases (SE < 1), the difference explains the level of inefficiency for the mentioned DMU. In line with this, a DMU will be scale efficient if SE score corresponds to 1.

10.4 Empirical Studies

10.4.1 Data and Methodology

In the empirical studies, TE, PTE, and SE values were measured and evaluated for the Turkish banks. In total, data regarding the 29 and 28 COMs and 12 INVs have been submitted to the DEA for 2010 and 2011. The data used in the DEA applications have been provided from the Bank Association of Turkey's official website (www.tbb.org.tr). Selecting the appropriate inputs and outputs has a critical role in the DEA process. In this respect, the appropriate inputs and outputs were chosen to reveal the real efficiency differences of DMUs.

The number of branches, number of personnel, equity capital, interest expenses, and noninterest expenses were the selected inputs. After considering the operation areas of the banks, loans, net profit, and deposits were selected as outputs for the COMs. However, loans and net profit values were chosen for the INVs.

The data regarding the COMs and INVs were submitted to the CCR model in accordance with Equation 10.2 and then had been analyzed by the BCC model in pursuance of Equation 10.5. In further, the SE scores of the Turkish banks had been determined according to

Equation 10.10. Regarding the analyses, Tables 10.2 through 10.5 are presented throughout the section.

10.4.2 Efficiency Results for Commercial Banks

In the first part of the DEA applications, the CCR and BCC models were applied for each COM, and TE, PTE, and SE scores were measured. The efficiency results regarding the COMs for 2010 and 2011 are given in Table 10.2.

As shown in Table 10.2, the CCR analyses reveal that six banks were efficient in 2010. The analyses also show that 23 banks have been found inefficient, while they have an efficiency level above 0.500. The banks have the average TE as 0.784. On the basis of the BCC model, 16 banks have PTEs, and the mean PTE level is 0.913.

Table 10.2 also shows the difference between TEs and PTEs, and six banks are found scale efficient for 2010. Besides, 23 banks are left scale inefficient, which means that these banks are not operating under the optimal scale.

According to Table 10.2, the CCR analyses show that nine banks were efficient in 2011. However, 19 banks have been found inefficient. The banks have the average TE as 0.808. In terms of BCC model, 17 banks have PTEs, and the mean PTE level is 0.915. Table 10.3 reveals the difference between TEs and PTEs, and 9 banks are found scale efficient for 2011, while 19 banks have been found scale inefficient.

10.4.3 Efficiency Results for Investment Banks

The CCR model and the BCC model had been applied for each investment bank for the second part of the DEA implementations for 2010 and 2011. The efficiency scores for the INVs are shown in Table 10.3.

As illustrated in Table 10.3, the CCR model results show that five banks were technically efficient in 2010. Six banks have been found inefficient. The average TE for the INVs is 0.674. According to the BCC model results, 10 banks have PTE, and the mean PTE level corresponds to 0.891. While five banks have shown SE and operate under the optimal scale, seven banks have shown scale inefficiency.

Table 10.2 Efficiency Results for Turkish Commercial Banks in 2010–2011

	2010			2011		
COMMERCIAL BANKS	TE	STE	SE	TE	STE	SE
ABN Amro Bank	0.775	1.000	0.775	1.000	1.000	1.000
Akbank	0.706	1.000	0.706	0.961	1.000	0.961
Alternatifbank	0.637	0.835	0.763	0.752	0.855	0.880
Anadolu Bank	0.677	0.857	0.791	0.758	0.838	0.905
Arap Turk Bank	1.000	1.000	1.000	1.000	1.000	1.000
Bank Mellat	1.000	1.000	1.000	1.000	1.000	1.000
Citibank	0.825	0.974	0.847	0.687	1.000	0.687
Denizbank	0.771	1.000	0.771	1.000	1.000	1.000
Deutsche Bank	1.000	1.000	1.000	1.000	1.000	1.000
EurobankTekfen	0.426	0.439	0.969	0.478	0.528	0.906
Finans Bank	0.643	0.837	0.769	0.650	0.767	0.847
Fortis Bank	0.530	0.765	0.693	N/A	N/A	N/A
Habib Bank Limited	1.000	1.000	1.000	1.000	1.000	1.000
HSBC Bank	0.730	0.912	0.801	0.672	0.891	0.754
ING Bank	0.665	1.000	0.665	0.835	1.000	0.835
Millennium Bank	0.967	1.000	0.967	1.000	1.000	1.000
Societe Generate	0.513	0.594	0.864	0.731	0.905	0.808
Şekerbank	0.729	0.801	0.910	0.717	0.859	0.834
Tekstil Bank	0.700	0.850	0.823	0.758	0.825	0.918
TEB	0.811	0.957	0.847	0.750	0.979	0.766
Turkish Bank	0.799	0.835	0.956	0.458	0.494	0.928
Turkland Bank	0.812	0.814	0.998	0.645	0.680	0.948
TCZB—Ziraat Bank	1.000	1.000	1.000	0.801	1.000	0.801
Turkish Garanti Bank	0.722	1.000	0.722	1.000	1.000	1.000
Turkish Halk Bank	0.957	1.000	0.957	1.000	1.000	1.000
Turkish İs Bank	0.783	1.000	0.783	0.809	1.000	0.809
Turkish Vakıflar Bank	0.787	1.000	0.787	0.949	1.000	0.949
West LB AG	1.000	1.000	1.000	0.243	1.000	0.243
Yapı Kredi Bank	0.783	1.000	0.783	0.962	1.000	0.962
Average	0.784	0.913	0.860	0.808	0.915	0.884

As we evaluate Table 10.3, the CCR model results reveal that 10 banks are technically efficient in 2011, while 2 banks have been found inefficient. The average TE score for the INVs is 0.988. On the basis of the BCC model, all banks have PTE, and the mean PTE level corresponds to 1. In this respect, 10 banks have shown SE and operate under the optimal scale, while 2 banks demonstrate scale inefficiency.

Table 10.3 Efficiency Results for Turkish Investment Banks in 2010–2011

	2010			2011		
INVESTMENT BANKS	TE	STE	SE	TE	STE	SE
Bank Pozitif Credit and Devp. Bank	0.323	0.331	0.977	0.889	1.000	0.889
Calyon Investment Bank Türk	1.000	1.000	1.000	1.000	1.000	1.000
Diler Investment Bank	1.000	1.000	1.000	1.000	1.000	1.000
GSD Investment Bank	0.476	1.000	0.476	0.971	1.000	0.971
Bank of Provinces	1.000	1.000	1.000	1.000	1.000	1.000
ISE Settlement and Custody Bank	0.332	1.000	0.332	1.000	1.000	1.000
Merrill Lynch Investment Bank	0.880	1.000	0.880	1.000	1.000	1.000
Nurol Investment Bank	0.210	0.357	0.588	1.000	1.000	1.000
Taib Investment Bank	0.310	1.000	0.310	1.000	1.000	1.000
Türk Eximbank	1.000	1.000	1.000	1.000	1.000	1.000
Turkish Development Bank	0.553	1.000	0.553	1.000	1.000	1.000
TSKB Industrial Development Bank of Turkey	1.000	1.000	1.000	1.000	1.000	1.000
Average	0.674	0.891	0.760	0.988	1.000	0.988

Thus, average SE score of INVs is almost 1. In addition, TE, PTE, and SE scores in 2010 are more efficient than that of 2011.

This shows that the global crisis did not adversely affected the Turkish banks in 2011 and that Turkish COMs and INVs increased their efficiencies in comparison to 2010.

10.4.4 Results for the Improvement Ratios

Tables 10.4 and 10.5 summarize the 2010–2011 results for the improvement ratios of the COMs and INVs for DEA models.

Table 10.4 reveals the average improvement ratios for the COMs in 2010. As per the results, inputs for the CCR model should be decreased by 1.32%–40.72%, and inputs for the BCC model have to be diminished by 0.43%–17.52%. The CCR model suggests improvement in outputs between 39.30% and 164.81%. Therefore, the BCC model indicates an output increase between 16.60% and 63.55%.

Table 10.4 shows the average improvement ratios for the COMs in 2011. According to the results, inputs for the CCR model should be decreased by 1.98%–18.85%, and inputs for the BCC model have to be diminished by 0.52%–11.41%. The CCR model suggests an

Table 10.4 Improvement Ratios for Turkish Commercial Banks in 2010–2011

	2010		2011	
DMUS	CCR	BCC	CCR	BCC
Number of branches	−0.366	−0.145	−0.188	−0.114
Number of personnel	−0.407	−0.175	−0.120	−0.108
Equity capital	−0.013	−0.004	−0.030	−0.005
Interest expenses	−0.018	−0.009	−0.020	−0.018
Noninterest expenses	−0.259	−0.119	−0.160	−0.083
Loans	0.393	0.166	0.488	0.132
Net profit	1.648	0.636	1.157	0.851
Deposits	0.418	0.236	0.858	0.319

improvement in outputs between 48.80% and 115.73%, while the BCC model indicates an output increase between 13.16% and 85.13%.

Table 10.5 illustrates the average improvement ratios for the INVs in 2010. Inputs for the CCR model have to be decreased by 11.90%–37.89%, and inputs for the BCC model should be diminished by 1.26%–9.68%. In addition, in terms of the DEA outputs, the CCR model suggests an improvement in outputs between 187.11% and 322.74%, and the BCC model points out an output increase between 69.77% and 82.72%. Moreover, Tables 10.4 and 10.5 illustrate the 2011 improvement ratios for the COMs and INVs.

Table 10.5 also shows the average improvement ratios for the INVs in 2011. According to the results, inputs for the CCR model should be diminished by 0.00%–5.49% and inputs for the BCC model do not suggest a decline. However, on the basis of outputs, the CCR model

Table 10.5 Improvement Ratios for Turkish Investment Banks in 2010–2011

	2010		2011	
DMUS	CCR	BCC	CCR	BCC
Number of branches	−0.262	−0.039	−0.029	0.000
Number of personnel	−0.135	−0.014	−0.055	0.000
Equity capital	−0.379	−0.050	0.000	0.000
Interest expenses	−0.119	−0.097	−0.042	0.000
Noninterest expenses	−0.194	0.013	−0.055	0.000
Loans	3.227	0.698	0.019	0.000
Net profit	1.871	0.827	0.836	0.000

suggests an improvement in outputs between 0.01% and 83.6%, whereas the BCC model does not indicate an output decrease.

Upon our evaluation of the average ratio results related to the COMs and INVs in Turkey between 2010 and 2011, we have reached amazing and crucial results. First of all, the results of the CCR model in 2010 regarding the outputs of the COMs and INVs came out enormously large.

Consequently, the net profit of the COMs should be increased from 21.373 billion TL to 56.597 billion TL. Moreover, the loans of the INVs indicate an improvement from 18.038 billion TL to 76.253 billion TL. However, the average ratio of inputs and outputs related to the COMs and INVs in 2011 decreased to that of 2010. As the COMs and INVs in Turkey examined the results in 2010, both efficiency scores and average ratios diminished in 2011.

Figure 10.1 illustrates the DEA efficiencies of the COMs and INVs in Turkey during the period between 2010 and 2011. As illustrated in Figure 10.1, the COMs demonstrated higher financial efficiency scores according to both DEA models in 2010, whereas the INVs showed higher financial efficiency scores consistent with the CCR and BCC models in 2011. The COM showed a 2% increase in efficiency in 2011 according to the CCR model. Moreover, the COM demonstrated a 0.2% improvement in efficiency in 2011 according to the BCC model.

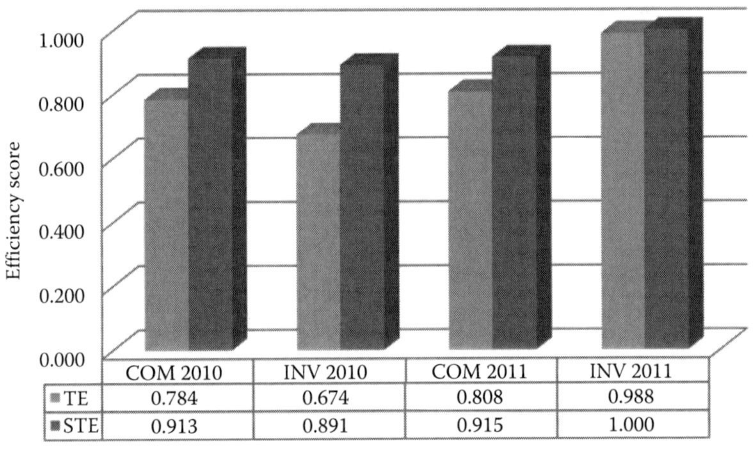

	COM 2010	INV 2010	COM 2011	INV 2011
TE	0.784	0.674	0.808	0.988
STE	0.913	0.891	0.915	1.000

Figure 10.1 Efficiency scores of commercial and investment banks. (Prepared by authors.)

In addition, the INV showed a 31.4% increase in efficiency in 2011 according to the CCR analysis, and a 10.9% improve according to the BCC analysis. Additionally, the average efficiency scores were also taken into consideration. Therefore,

- 50.00% of COM according to the CCR model and 83.33% of INV according to the CCR model in 2011
- 44.83% of COM according to the CCR model and 50.00% of INV according to the CCR model in 2010
- 64.29% of COM according to the BCC model and 100.00% of INV according to the BCC model in 2011
- 62.07% of COM according to the BCC model and 83.33% of INV according to the BCC model in 2010

showed higher performances in comparison to the average efficiency score. Therefore, the COMs and INVs in 2011 had a better average efficiency score than they had in 2010.

10.5 Conclusions

This study covers the efficiency analysis of the Turkish banks by the DEA method, which is a widely used mathematical programming technique. The empirical study has shown valuable efficiency results for the Turkish COMs and INVs in 2010 and 2011. In this framework, the CCR efficiency levels of INVs were found to have decreased in 2010 in comparison to the 2011 efficiency results, whereas the BCC efficiencies were unchanged.

On the other hand, the CCR and BCC efficiencies of the COMs diminished in the 2011 analyses as compared to the 2010 efficiency results. It is of great importance to achieve efficient frontiers for the DMUs, and hence the improvement ratios for each input and output should be considered. According to the results regarding the improvement ratios for the DEA models, mostly the output values of the COMs required smaller improvements, while the outputs of the INVs needed higher improvements.

However, the inputs of the COMs and INVs should be improved in similar amounts on the basis of both DEA models. In general, it is surprising that the INVs had superior financial efficiencies in comparison to the COMs in 2010 and 2011. This may depend upon the progress in the real sector for the mentioned period. Further, the analyses have shown that there are still inefficiencies within the COMs and INVs in Turkey. Thus, the Turkish banks should improve their operating figures to increase efficiencies in order to compete within the global financial markets. Consequently, the DEA is a valuable mathematical programming tool in analyzing financial efficiencies of the Turkish COMs and INVs.

References

Banker, R. D., Charnes, A., and Cooper, W. W. 1984. Some models for estimating technical and scale inefficiencies in data envelopment analysis. *Management Science*. 30: 1078–1092.

Barrientos, A. and Boussofiane, A. 2005. How efficient are pension funds in Chile? *R. Econonmia Contemporanea*. 9(2): 289–311.

Baurer, P. W. 1993. Efficiency and technical progress in check processing. *Economic Review*. 3: 24–38.

Berger, A. N. and Humprey, D. B. 1997. Efficiency of financial institutions: International survey and directions for future research. *European Journal of Operational Research*. 98: 175–212.

Berger, A. N., Hunter, W. C., and Timme, S. G. 1993. The efficiency of financial institutions: A review and preview of research past, present and future. *Journal of Banking and Finance*. 17: 221–249.

BRSA. 2006. Financial markets report (December). Banking Regulation and Supervision Agency of Turkey, Ankara, Turkey.

BRSA. 2007. Financial markets report (December). Banking Regulation and Supervision Agency of Turkey, Ankara, Turkey.

BRSA. 2009. Financial markets report (December). Banking Regulation and Supervision Agency of Turkey, Ankara, Turkey.

BRSA. 2010. From crisis to financial stability (September). Banking Regulation and Supervision Agency of Turkey, Ankara, Turkey.

BRSA. 2011. Financial markets report (December). Banking Regulation and Supervision Agency of Turkey, Ankara, Turkey.

Busso, A. and Funari, S. 2001. A data envelopment analysis approach to measure the mutual funds performance. *European Journal of Operational Research*. 135: 477–492.

Carbo, S., Gardener, E. P. M., and Williams, J. 2002. Efficiency in banking: Empirical evidence from the savings bank sector. *The Manchester School*. 70: 204–228.

Charnes, A., Cooper, W. W., and Rhodes, E. 1978. Measuring the efficiency of decision making units. *European Journal of Operational Research*. 98: 408–418.

Choi, Y. K. and Murthi, B. P. S. 2001. Relative performance evaluation of mutual funds: A non-parametric approach. *Journal of Business Finance & Accounting*. 28(7): 853–876.

Çingi, S. and Tarım, A. 2000. Performance measurement in Turkish banking system: DEA malmquist TFP index application [Türk banka sisteminde performans ölçümü: DEA malmquist TFP endeksi uygulaması]. *Turkish Banks Association Research Proceedings Series*. 2: 1–34.

Conceicao, A. M., Portela, S., and Thanassoulis, E. 2007. Comparative efficiency analysis of Portuguese bank branches. *European Journal of Operational Research*. 177(2): 1275–1288.

Cooper, A., Seiford, L. M., and Tone, K. 2000. *Data Envelopment Analysis*. Kluwer Academic Publishers, Norwell, MA, pp. 1–306.

Cooper, A., Seiford, L. M., and Tone, K. 2006. *Data Envelopment Analysis and Its Uses*. Springer, Berlin, Germany, pp. 1–342.

Ertuğrul, A. and Zaim, O. 1999. Economic crises and efficiency in Turkish banking industry. *METU Studies in Development*. 26(1–2): 99–116.

Fandel, P. 2003. Technical and scale efficiency of corporate farms in Slovakia. *Agricultural Economics—Czech*. 49: 375–383.

Farrell, M.J. 1957. The measurement of productive efficiency. *Journal of the Royal Statistic Society*. A120: 253–281.

Gökgöz, F. 2009a. Data envelopment analysis and its application to finance area [Veri zarflama analizi ve finans alanına uygulanması] Ankara University Faculty of Political Sciences Publication. No: 597.

Gökgöz, F. 2009b. Data envelopment analysis for Turkish banks: Evidence on the financial efficiencies of the commercial and investment banks. *Banking and Finance Letters*. 1(2): 43–50.

Gökgöz, F. 2010. Measuring the financial efficiencies and performances of Turkish funds. *Acta Oeconomica*. 60(3): 295–320.

Karacabey, A. A. 2002. A quantitative study on productivity changes in the Turkish banking sector. *İktisat İşletme ve Finans Review*. 191: 68–78.

Kasman, A. 2002. Cost efficiency, scale economies and technological progress in Turkish banking. *Central Bank Review*. 1: 1–20.

Koopmans, T. C. 1951. An analysis of production as an efficient combination of activities, in: Koopmans T.C. (Ed.), *Activity Analysis of Production and Allocation*, Cowles Commission for Research in Economics, Monograph No: 13. Wiley, New York, pp. 33–97.

Mercan, M. and Yolalan, R. 2000. The effect of scale and mode of ownership on the Turkish banking sector financial performance. *Istanbul Stock Exchange Review*. 4(15): 2–23.

Murthi, B. P. S., Choi, Y. K., and Desai, P. 1997. Efficiency of mutual funds and portfolio performance measurement: A non-parametric approach. *European Journal of Operational Research*. 98: 408–418.

Ruggiero, J. 2000. Measuring technical efficiency. *European Journal of Operational Research*. 121: 138–150.

Zaim, O. 1995. The effect of financial liberalization on the efficiency of Turkish commercial banks. *Applied Financial Economics*. 5: 257–264.

Online Resources

SDIF. 2003. Annual report. Turkish Savings Deposit Insurance Fund. http://www.tmsf.org.tr/yillik.rapor.en (accessed December 9, 2013).

TBB. 2011. Banks Association of Turkey. http://www.tbb.org.tr/tr/banka-ve-sektor-bilgileri/istatistiki-raporlar/—2011—secilmis-rasyolar/1172 (accessed December 9, 2013).

Treasury. 2013. Turkish economy presentation. http://www.treasury.gov.tr/default.aspx?nsw=QQl5h+br0ZErvDuyH2bTyg==-SgKWD+pQItw=&nm=853. (accessed December 10, 2013)

Turkish Statistical Institute. http://www.turkstat.gov.tr/PreIstatistikTablo.do?istab_id=1526 (accessed December 10, 2013).

Author Index

A

B

C

J

K

L

M

N

O

P

Q

Subject Index

D

E